Y. Morishima · T. Norisuye · K. Tashiro (Eds.)

Molecular Interactions and Time-Space Organization in Macromolecular Systems

Springer
Berlin
Heidelberg
New York
Barcelona
Hong Kong
London
Milan
Paris
Singapore
Tokyo

Y. Morishima · T. Norisuye · K. Tashiro (Eds.)

Molecular Interactions and Time-Space Organization in Macromolecular Systems

Proceedings of the OUMS '98, Osaka, Japan, 3-6 June, 1998

With 94 Figures and 11 Tables

Springer

Editors

Professor Yotaro Morishima
Professor Takashi Norisuye
Professor Kohji Tashiro

Osaka University
Graduate School of Science
Department of Macromolecular Science
Toyonaka, Osaka 560-0043
Japan

ISBN 3-540-66110-7 Springer-Verlag Berlin Heidelberg New York

Library of Congress Cataloging-in-Publication Data
OUMS '98 (3rd : 1998 : Osaka, Japan)
Molecular interactions and time-space organization in macromolecular systems : proceedings of the OUMS '98, Osaka, Japan, 3-6 June, 1998 / Yotaro Morishima, Takashi Norisuye, Kohji Tashiro, eds.
Includes bibliographical references.
ISBN 3-540-66110-7 (hardcover : alk. paper)
1. Macromolecules Congresses. 2. Molecular dynamics Congresses.
I. Morishima, Yotaro. II. Norisuye, T. (Takashi) III. Tashiro, Kohji. IV. Title
QD380.094 1998547'.7--dc21 99-34982

Typesetting: Camera-ready by editors
Cover layout: Struve & Partner, Heidelberg
SPIN: 10684898 02 / 3020 - 5 4 3 2 1 0 - Printed on acid-free paper

Preface

This book is a collection of lectures presented at the third Osaka University Macromolecular Symposium (OUMS '98) on "Molecular Interactions and Time-Space Organization in Macromolecular Systems" which was held at the Osaka University Convention Center, Osaka, Japan, on June 3-6, 1998. The symposium was sponsored by the Osaka University International Exchange Program in Macromolecular Science and supported in part by the Ministry of Education, Science, Sports, and Culture, Japan and by the Society of Polymer Science, Japan. The Osaka University Macromolecular Symposium is intended to be an international meeting on the state of the art in macromolecular science bringing together scientists from all parts of the world who are experts in macromolecular chemistry, physical chemistry, and physics.

Progress in polymer science in the 60s and 70s took place mainly by studies of relatively simple systems such as dilute polymer solutions and polymer crystals. However, in the last decades, growing attention has been focused on more complicated systems and phenomena in which complicated molecular interactions are involved as a key element.

The OUMS '98 highlighted recent advances in studies of macromolecular interactions, organization, and dynamics in systems of crystalline polymers, liquid crystalline polymers, glassy polymers, block copolymers, polyelectrolytes, amphiphilic polymers, and polymer assemblies. This area of macromolecular science has enjoyed tremendous growth over the last decade. With the key words of "molecular interactions" and "molecular organization", we expanded the scope of the symposium from purely macromolecular systems to the more general area of molecular assemblies and polymer interactions with micelles, liposomes, and proteins.

At the symposium, 24 invited lectures were presented, and 20 lectures contributed to this book. The lectures included experimental results and methods, theory, and computer simulations. Besides the invited lectures, many interesting new results were presented by the participants in the poster sessions during the meeting. Although many presentations were concerned with basic problems in macromolecular science, we have not excluded practical applications of macromolecular systems that are somehow related to the symposium topics.

It is our belief that this book is certain to be useful to those who are actively involved in the research in this related area of macromolecular science and also to those who wish to get familiar with these areas because the papers presented in this book are of top quality not available collectively in any other publication.

We close this preface with sincere thanks to the contributors for their extended cooperation and to the staff of Springer for professional guidance for editing this book.

March 1999

Yotaro Morishima
Takashi Norisuye
Kohji Tashiro

Table of Contents

Synchrotron SAXS and WAXD Study of Polymer Crystallization in Confined Microdomains

Li-Zhi Liu and Benjamin Chu

Department of Chemistry, State University of New York at Stony Brook
Stony Brook, Long Island, New York 11794-3400

Introduction

Crystallization of polymers in confined microdomains could exhibit different features, including crystalline structure, morphology and novel dynamic processes. Two different such systems have been investigated. They are crystallizable mixtures of a block copolymer with a corresponding homopolymer and miscible semicrystalline/semicrystalline homopolymer blends. In a compatible blend of an AB diblock copolymer and its corresponding homopolymer A, spherical, cylindrical, lamellar, hexagonally-perforated (catenoid) lamellar, bicontinuous I*ad* cubic (gyroid) and bicontinuous P*nm* cubic (double diamond) microphases[1-4] can be formed. In AB/A compatible blends with one of the two blocks crystallizable, the crystallization behavior, the structure and the morphology could be very different from the extensively studied miscible polymer blends [5,6]. In this lecture, the AB/A blends with "hard" amorphous microdomains, meaning that the glass transition temperature, Tg, of the domains is higher than the melting point, Tm, of the crystallizable block, has been studied. Schematic diagrams of the systems are shown in Figure 1. Miscible semicrystalline/semicrystalline blends

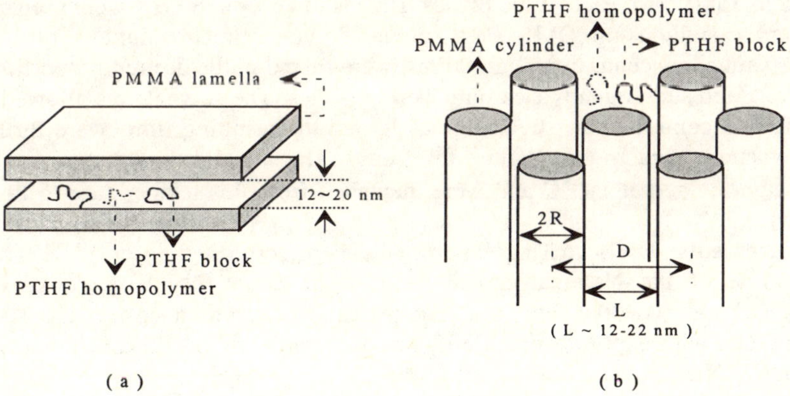

Fig. 1. Schematic diagrams of blend systems with different microphase separation structures. (a) alternating PTHF and PMMA lamellae; (b) PTHF matrix with PMMA dispersed as cylinders arranged in hexagonal form (the definitions R, L and D are also shown in this figure). Reprinted from Reference 6.

Y. Morishima · T. Norisuye · K. Tashiro (Eds.)
Molecular Interactions and Time-Space Organization in Macromolecular Systems
Proceedings of the OUMS '98, Osaka, Japan, 3-6 June, 1998

can also exhibit a complicated phase behavior. The crystallization of the component with a lower melting point can occur in a confined lamellar region. Compatible mixtures of tetrahydrofuran-methyl methacrylate diblock copolymers (PTHF-b-PMMA) with tetrahydrofuran homopolymer (PTHF) [5-6], and semicrystalline/semicrystalline miscible blends of poly(vinylidene fluoride) (PVF_2) and poly(1, 4-butylene adipate) (PBA) [7] were studied with synchrotron X-rays.

Experiment

Materials. The synthesis of tetrahydrofuran-methyl methacrylate diblock copolymers (PTHF-b-PMMA) and the purification have been described elsewhere [8]. A summary on the pertinent parameters of the PMMA-b-PTHF diblock copolymers and the PTHF homopolymer used in the present work is given in Table 1. The PVF_2 ($M_w=1.4\times10^5$) sample used was supplied by Polysciences Inc. (Warrington, PA) and was used as received. The PBA sample was obtained from Scientific Polymer Products, Inc. (Ontario, NY) and was purified by reprecipitation from N, N-dimethylformamide (DMF) into cold methanol (-10 °C). The weight average and number average molecular weights of the as-precipitated PBA were 1.4×10^4 and 9.8×10^3, respectively, as determined by gel permeation chromatography using 1, 4-dioxane as the eluent.

Preparation of Blends. Blends of the PTHF-b-PMMA copolymer and a low-molecular-weight homopolymer PTHF were prepared by solution casting from chloroform. Compositions of the blends, including W(H-PTHF) denoting the PTHF homopolymer weight fraction in the blend, W(THF in blend) denoting the total weight fraction of THF of both PTHF block and PTHF homopolymer in the blend, and W(H-PTHF in THF phase) denoting the weight fraction of PTHF homopolymer in the THF phase under the assumption of strong segregation limit are listed in Table 2 for each of the blends. Blends of PVF_2 and PBA were prepared by solvent casting using DMF as a mutual solvent. Stock solutions of both polymers (initial concentration: 2 g/100 ml) were mixed in the desired proportions, well-stirred, and subsequently cast onto Teflon dishes. The solvent was allowed to evaporate in a controlled air stream for 24 h, and the resulting films were further dried in vacuum at room temperature for 3 days. The samples thus prepared were compression molded at 190°C and were quenched from this temperature to room temperature.

Measurements. The scattering experiments were performed at the SUNY X3-A2 beamline of the National Synchrotron Light Source(NSLS), Brookhaven National Laboratory(BNL). Fuji imaging plates were used to collect the static scattering data with exposure times of 2 min per frame for SAXS and 1 min per frame for WAXD.

Table 1. Characterization of PTHF-b-PMMA diblock copolymers and PTHF homopolymer. Reprinted from Reference 6.

Sample[a]	$M_n^b \times 10^{-3}$	W^c(PTHF)	M_n (PTHF Block)$\times 10^{-3}$	M_n (PMMA block)$\times 10^{-3}$
1TM	23.3	0.30	7.0	16.3
2TM	14.0	0.50	7.0	7.0
PTHF homopolymer			2.0	

a, T and M denote abbreviations PTHF and PMMA, respectively;
b, M_n: number average molecular weight.
c, W: weight fraction

Table 2. Compositions of PTHF-b-PMMA/PTHF Blends. Reprinted from Reference 6.

Designation	W(H-PTHF)	W(THF in blend)*	W(H-PTHF in THF phase)
1TM/T1	0.1	0.37	0.27
1TM/T2	0.2	0.44	0.45
1TM/T3	0.3	0.51	0.59
1TM/T4	0.4	0.58	0.69
1TM/T5	0.5	0.65	0.77
2TM/T2	0.2	0.60	0.33
2TM/T3	0.3	0.65	0.46
2TM/T4	0.4	0.70	0.57

* including PTHF homopolymer and PTHF block of the copolymer

Results and Discussion

1. Crystallization and morphology in PTHF-b-PMMA/PTHF blends

WAXD study of crystalline structure in PTHF microdomains. The microphase morphology for each individual blend in the melt-state (above the melting temperature of PTHF) was determined on the basis of scattering peak positions from SAXS and is listed in Table 3. Our SAXS study showed that when the samples with PMMA lamellae or cylinders in the melt state were cooled down to crystallize, the microphase morphology and interdomain distance remained unchanged, because the glass transition temperature of PMMA lamellae or cylinders was much higher than the melting point of the PTHF component. Therefore, the crystallization of the PTHF block and the PTHF homopolymer occurs in the existing PTHF microdomain. WAXD patterns of the blends 1TM/T1 to 1TM/T5, as well as of the pattern of neat copolymer 1TM, are shown in Figure 2. From this figure, the following diffraction features were observed. (1) No diffraction peak was seen for the pure copolymer 1TM. (2) Two sharp diffraction peaks were seen for the 1TM/T1 blend with as little as 10 wt % of PTHF homopolymer. (3) The diffraction peak position shifts to a higher θ-range with an increase in the PTHF homopolymer content of the blend. (4) The full width at half-maximum intensity of the diffraction peak at $2\theta \sim 19.7°$ remained almost unchanged with an increase in the homopolymer weight fraction in the blends. However, the full width at half-maximum intensity of the diffraction peak at $2\theta \sim 24°$ decreased with an increase in the PTHF homopolymer content of the blend. (5) The diffraction intensity of (202) peak increased progressively with increasing PTHF homopolymer content, while the diffraction intensity of (002) peak showed a very large increase when the homopolymer content was increased from 10 to 20 wt %, and then remained almost constant with further increase in the homopolymer content. According to the above diffraction features, the suggested chain-folding model in the confined lamella had a fiber identity along the normals of the PMMA lamella.

Table 3. Microphase morphology of PTHF-b-PMMA/PTHF blends studied. Reprinted from Reference 6.

Designation	W (THF in blend)*	Microphase morphology
1TM/T1	0.37	Lamella
1TM/T2	0.44	Lamella
1TM/T3	0.51	Lamella
1TM/T4	0.58	PMMA cylinder
1TM/T5	0.65	PMMA cylinder
2TM/T2	0.60	PMMA cylinder
2TM/T3	0.65	PMMA cylinder
2TM/T4	0.70	PMMA cylinder

* including PTHF homopolymer and PTHF block of the copolymer

Crystallization in PTHF matrix dispersed with PMMA cylinders. In the present work, some new scattering features have been observed for a series of blends with hexagonally packed PMMA cylinders. The higher-order scattering peaks seemed to shift to lower q values after crystallization, where q ($=(4\pi/\lambda)\sin\theta$) is the magnitude of the scattering wave vector. The $(4)^{1/2}$ higher order peak disappeared and the $(3)^{1/2}$ higher order peak appeared at a lower q range, as shown in Figure 3. The $(7)^{1/2}$ higher order peak seemed to have shifted to a lower q value. By comparing the inset figure in Figure 3 and Figure 4 for the scattering profiles on a linear scale, the scattering contribution from PTHF crystals was much larger in the

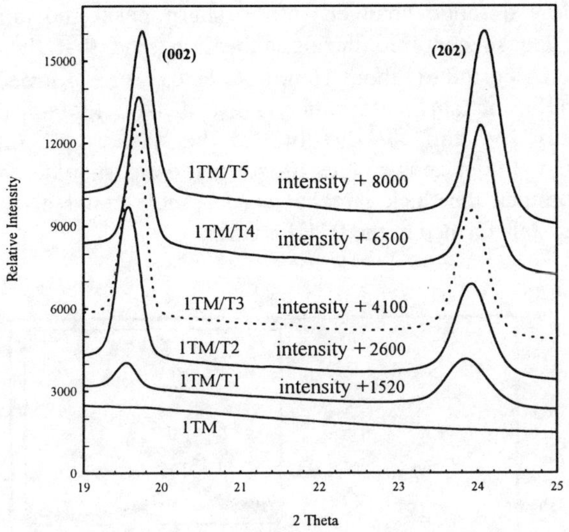

Fig. 2. WAXD patterns of 1TM pure copolymer and of its blends with 10 to 50 wt % of PTHF homopolymer. 2θ is the scattering angle. Reprinted from Reference 6.

blend 2TM/T4 than in the blend 2TM/T3, due to the higher PTHF homopolymer content in the former sample, which increased the intersurface distance L between two neighboring PMMA cylinders (see Fig 1). After crystallization, two large scattering shoulders at q ~ 0.4 and 0.5 nm^{-1} were observed (Fig. 4), indicating the formation of ordered PTHF lamellar stacks. After annealing for a month, the

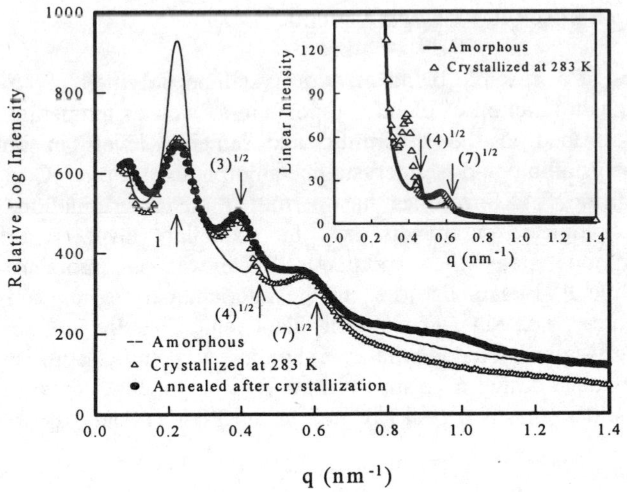

Fig. 3. SAXS profiles of the blend 2TM/T3. Reprinted from Reference 6.

shoulder at a low q range changed into a sharp peak, indicating that further ordering of the stacks occurred during annealing, and that the PTHF lamellar stacks with a long period of about 16 nm ($\approx 2\pi/q_m$) were formed in the matrix microphase. As the intersurface distance L of PMMA cylinders (Figure 1) for the 2TM/T4 was only 15.8 nm, the direction of the stack could only be along the direction of the cylinder axis. The broad scattering shoulder at q ~ 0.5 nm^{-1} suggested that besides the thick lamellar stacks, there could also be some PTHF irregular microcrystals formed in the PTHF matrix.

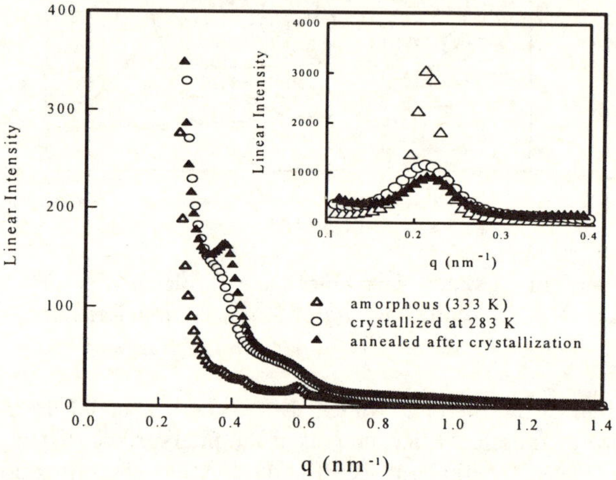

Fig. 4. SAXS profiles of the blend 2TM/T4. Reprinted from Reference 6.

2. Crystallization in a miscible PVF_2/PBA blend

The morphologies of a miscible blend of two crystalline polymers (PVF_2/PBA) have been examined as a function of blend composition as well as temperature. The morphologies were probed at the spherulitic and lamellar levels in both the semicrystalline/semicrystalline and semicrystalline/amorphous states. Correlation function analysis of the SAXS profiles has permitted the determination of the characteristic morphological parameters at the lamellar level. For the semicrystalline/amorphous state, the amorphous PBA was incorporated partly between the crystalline PVF_2 lamellae, i.e., in the interlamellar region, and partly within the spherulites, i.e., in the interfibrillar region. In the case of semicrystalline/semicrystalline blends, the morphology was more complex. In addition to the pronounced scattering peak from the PVF_2 crystalline lamellae, there was a scattering shoulder (or small peak) related to the PBA crystals, as shown in Figure 5,

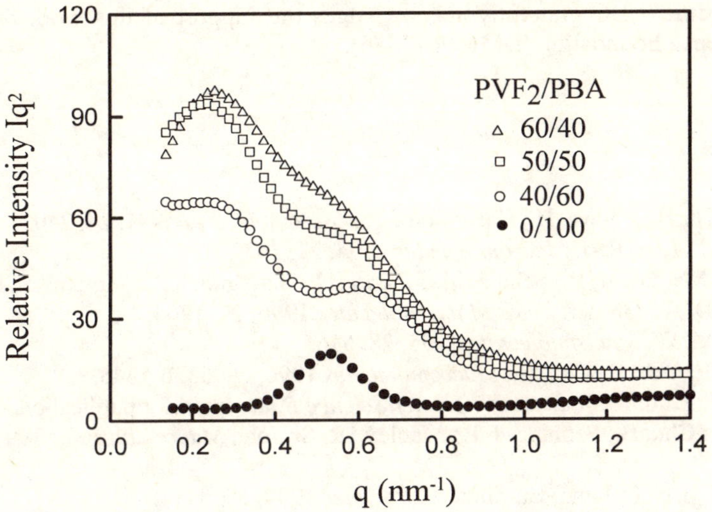

Fig. 5. Lorentz-corrected SAXS profiles of PBA homopolymer and the PVF_2/PBA blends with 40 - 60 wt% of PBA in the semicrystalline/semicrystalline state (~20°C). Reprinted from Reference 7.

indicating that the PVF_2 and PBA lamellae were in a segregated arrangement. Taken together, the data led to the conclusion that PBA crystallized as a thick lamella in the PVF_2 interlamellar region, and that there was a mixed amorphous phase on either side of the PBA lamellae, i.e., between the PBA and PVF_2 crystal lamellae. This structure model for the semicrystalline/semicrystalline PVF_2/PBA blends was different from the model proposed by Cheung *et al.* [9] for semicrystalline/semicrystalline (PCL/PC) blends, namely a random mixture of PC and PCL lamellae.

Conclusions

For the compatible PTHF-b-PMMA/PTHF blends, the x-ray diffraction study showed that the crystalline structure in a confined PTHF lamella was very sensitive to the thickness of the lamella. The SAXS study showed that when the interdistance of two neighboring PMMA cylinders was comparable to the PTHF long period, ordered PTHF stacks could be formed in the PTHF matrix. For the miscible semicrystalline/semicrystalline PVF_2/PBA blends, our results showed that only when the PVF_2 interlamellar distance was larger than the long period of the pure PBA homopolymer, the PBA component could crystallize as a thick lamella in the PVF_2 interlamellar region.

Acknowledgement BC gratefully acknowledges the support of this work by the National Science Foundation (DMR 9612386).

References

[1] Tanaka, H.; Hasegawa, H.; Hashimoto, T. *Macromolecules* 1991, 24, 240.
[2] Jeon, K.-J.; Roe, R.-J. *Macromolecules* 1994, 27, 2439.
[3] Zhao, J.; Majumdar, B.; Schulz, M. F.; Bates, F. S.; Almdal, K.; Mortensen, K.; Hajduk, D. A.; Gruner, S. M. *Macromolecules* 1996, 29, 1204.
[4] Matsen, M. W. *Macromolecules* 1995, 28, 5765
[5] Liu, L.-Z.;, Yeh, F., Chu, B., *Macromolecules* 1996, 29, 5336-5345.
[6] Liu, L.-Z.; Chu, B., *Journal of Polymer Science B*, accepted for publication.
[7] Liu, L-Z.; Chu, B.; Penning, J. P.; Manley, R. St. John *Macromolecules*,1997,30, 4398.
[8] Liu, L.-Z.; Li, H.; Jiang, B.; Zhou E., *Polymer* 1994, 35, 5511.
[9] Cheung, Y. W.; Stein, R. S.; Lin, J. S.; Wignall, G. D. *Macromolecules* 1994, 27, 2520.

Structure Development in Side Group Liquid Crystalline Diblock Copolymers

Chinedum O. Osuji, John T. Chen , Guoping Mao, Christopher K. Ober, Edwin L. Thomas

Department of Materials Science and Engineering, Massachusetts Institute of Technology, Cambridge MA 02139
Elt@mit.edu

Abstract. In side-chain liquid crystalline diblock polymers, driving forces for ordering of the material may be provided by the chemical incompatibility between the blocks and the liquid crystalline nature of one of the blocks. We study the microstructure and its development from initially isotropic solutions in side group liquid crystalline copolymers based on styrene-isoprene diblocks. Hierarchical structure from the 5 Å to the 500 Å length scale is observed, the product of coherent block copolymer microphase separation and liquid crystalline mesophase formation. The development of structure in these materials is discussed

1 Background

Liquid crystalline block copolymers (LC BCP's) are a relatively new class of polymers, of commercial interest because of their broad range of applications, and of academic interest since they permit the study of structure formation under the influence of more than one driving force. The physics of their ordering is analogous to that in crystallizable block copolymers such as poly(ethylene-b-styrene) or poly(ethylene-b-3-methyl-1-butene) [1] The formation of a liquid crystalline mesophase by one block typically requires the presence of a fairly mobile mesogenic group in that block. Incorporation of liquid crystallinity into block copolymers is possible via various synthetic routes, and many different chain architectures such as segmented main-chain, rod-coil and side group liquid crystalline block copolymers may be realized [2].

LC BCP's inherently possess a hierarchy of structure, with order present on the 10, 100 and 1000 Å length scales, corresponding to the inter-mesogen, smectic layer and microdomain spacings respectively. The manipulation of the chemistry and processing of liquid crystals or block copolymers to tailor their properties in some prescribed manner is a desirable goal. The driving forces for mesophase formation and microphase separation may also be similarly manipulated due to the tandem interaction of liquid crystallinity with microphase separation. Interest in these materials is growing quickly, as their liquid crystallinity, coupled to their ability to self-assemble, makes them attractive for application to sensing and display devices based on their electro-optic, mechano-optic, and thermo-optic properties. Piezo-, pyro- and ferro-electric responses, for instance, can be elicited from these materials [3, 4, 5, 6, 7] .

The concept of a liquid crystalline block copolymer was first fully realized by the synthesis by Percec *et al.* of poly(para-vinyl benzene ether-b-biphenyl acrylate) and poly(para-vinyl benzene-b-4-methoxy-4'-(ethoxy-(2-ethoxy-(2-ethoxy))) biphenyl acrylate) based copolymers [8]. Shortly thereafter, Adams and Gronski produced a styrene-butadiene diblock copolymer in which the butadiene block was side-functionalized with a cholesteryl mesogen [9]. The aim of their synthesis was to prepare liquid crystalline phases of well defined shape and size so that the influence of the size of the phase on the LC behavior as well as the effect of the interface on the LC order in the boundary region could be investigated [10].

A systematic study of the influence of liquid crystallinity on microdomain structure and vice-versa was conducted by Fischer *et al.* who produced a series of copolymers based on a cholesteryl functionalized ethyl methacrylate (PEMA) blocked first with styrene (PS) [11, 12] then later with butyl methacrylate (PBMA)[13]. The homopolymer formed using the cholesteryl functionalized methyl methacrylate monomer was smectic A. Using TEM and SAXS, in both systems they observed lamellae, spheres and cylinders of both PS and PBMA and spheres of the liquid crystalline block. Notably, the occurrence of cylinders of the LC component was *not* observed. In addition, a clear dependence of the mesophase type on the microphase separated structure was found. Specifically, the LC was nematic when confined to the discontinuous spherical domains, and smectic A otherwise. Clearly, the block copolymer morphology on the 100 to 1000 Å length scale was influencing the LC morphology on the smaller 10 to 40 Å length scale. The anchoring condition of the mesogen with respect to the IMDS was not determined, nor was the influence of spacer length explored. The noted absence of the cylindrical LC phase, as one might expect, is specific to the system and should not be universally expected [14, 2].

More recently, Hammond *et al.* [15] have synthesized a diblock copolymer based on styrene-methyl methacrylate in which the methyl methacrylate block is functionalized with a chiral mesogen using spacers of 6 and 10 carbon atoms. The mesogen in this case is based on a biphenyl benzoate core. Here coincidence is found between the order-disorder transition (ODT) and the LC isotropization temperature for block copolymers with high LC content and low molecular weight

[6]. The situation is analogous to that of crystallizable block copolymers in the weak segregation limit in which phase separation in driven by block crystallization. Currently, the synthesis of LC functionalized siloxane block copolymers is being pursued in an attempt to realize a smectic C* phase at room temperature. This has been successfully accomplished for the siloxane LC homopolymer already, and recently reported for a series of block copolymers[17].

Work conducted by the Ober and Thomas groups has focused on both LC rod-amorphous coil block copolymers[18, 19, 20, 21] and amorphous side group LC diblock copolymers [18, 19, 5, 22, 14], as well as fluorinated side group LC copolymers. For the amorphous non-fluorinated materials, two different mesogens were produced: one a chiral mesogen based on a biphenyl benzoate and the other an achiral mesogen having an azobenzene unit at its core. Both mesogens were used individually to produce two series of side group LC diblock copolymers employing isoprene to host the LC with styrene as the other block. The chemical structures of the styrene-isoprene backbone and the mesogens used appear in Figure 1. For the azobenzene series of polymers, using TEM and SAXS, microdomain structures of cylinders of poly(styrene), cylinders of poly(isoprene-LC) as well as lamellae were observed. Notably, spheres were not observed for the samples for which Φ_{LC}=0.13 and Φ_{LC} = 0.91, volume fractions at which

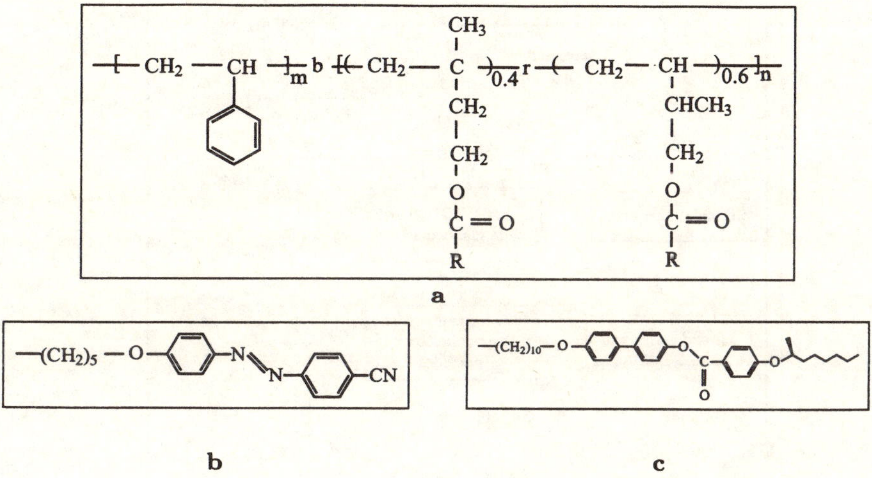

Fig. 1. P(S-b-ILC) materials. **a** Styrene-Isoprene backbone. **b** Achiral azobenzene derivative mesogen **c** Chiral biphenyl benzoate derivative mesogen

spheres are typically the most stable microdomain form in coil-coil diblock copolymers. The mesophase was smectic A in all cases for this series, and the mesogens were found, by x-ray diffraction from oriented samples, to anchor parallel to the inter-material dividing surface or IMDS.

2 Structure Formation

The development of order in LC BCP's must occur subject to constraints imposed by both the liquid crystalline mesophase and the block copolymer microdomain morphology. Ideally, structure formation is a coherent process, but it is probably often frustrated in practice. The equilibrium structure presented by a liquid crystalline block copolymer may be understood by giving consideration to the critical issues outlined below.

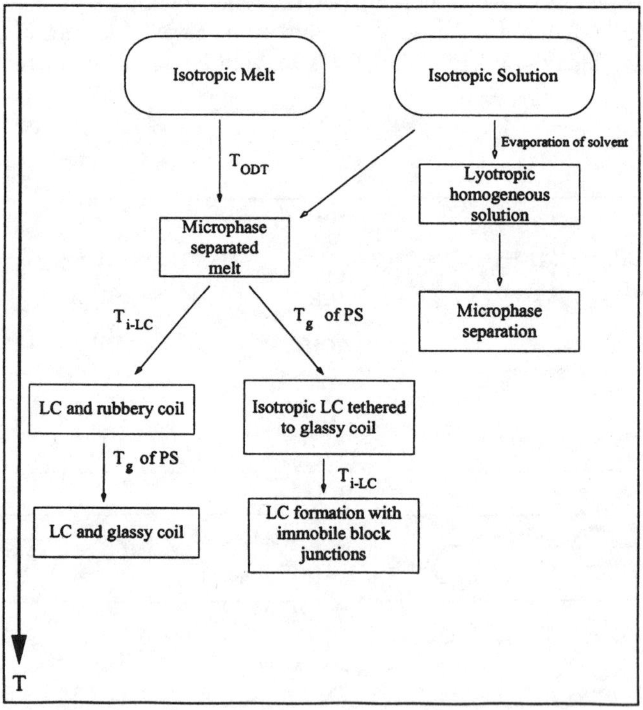

Fig. 2 Paths to final structure in P(S-b-ILC) LC BCP's

- Relative positions of thermal transition temperatures
- Preferred anchoring condition of mesogens at IMDS
- Volume fraction of block constituted by mesogen
- Curvature of IMDS
- Size of microdomain relative to size of mesophase
- Strength and density of LC defects
- Symmetry of native mesophase and expected microphase separated structure

The location of thermal transitions and the preferred anchoring condition of the mesogens at the IMDS are particularly important, as these characteristics influence the path dependence and the packing of the mesogens into their confining volumes respectively.

2.1 Influence of thermal character

The thermal transitions of interest in LC BCP's are the block copolymer order to disorder transition at T_{ODT}, the LC mesophase structural transitions such as at $T_{Sm(A)-N}$, T_{N-I} and $T_{Sm(A)-I}$ and the glass transitions of both blocks at T_g^A and T_g^B. Typically, because of the large chemical incompatibility between the blocks due to the presence of the mesogen, T_{ODT} for LC BCP's is quite high and thus inaccessible without thermal degradation of the material. In systems with very low molecular weight, T_{ODT} may actually be accessible, and one can witness microphase separation driven expressly by the formation of LC mesophases on passage through T_{I-LC} [6]. In this case, the morphology which forms ultimately will depend on whether the microphase separation that accompanies liquid crystallization can move the LC mesophases around in space to create periodically spaced microdomains. In crystallizable block copolymers, for instance, it is usually the case that crystallization induced phase separation results in the production of disordered microdomain structures [16, 1]. In the more general case, structure formation from the isotropic melt of an LC BCP above T_g^A and T_g^B as well as T_{I-LC} may proceed by two routes, following microphase separation on passage through the ODT, as depicted in Figure 2. If $T_{I-LC} > T_g^A$ and T_g^B then liquid crystallinity develops in a rubbery host block (A) which is tethered to another rubbery block (B) which then vitrifies. If $T_{I-LC} < T_g^A$, then mesophase formation is more difficult due to the decreased mobility of the mesogens in the glassy environment of the host coil. The degree to which the mobility of the mesogen is decreased of course depends on the volume fraction of the block that is occupied by the mesogens themselves, the architecture of the host block, and the temperature. If $T_{I-LC} < T_g^B$ and $T_{I-LC} < T_g^A$ then mesophase formation is even more difficult since the host backbone mobility is decreased not only by its glassy nature, but also by that of the block to which it is physically attached.

The evaporation of solvent from initially isotropic solutions may result in the formation of different morphologies depending, of course, on the solvent used.

The solvent may affect the final morphology by either encouraging or suppressing the formation of a lyotropic mesophase before microphase separation, which would occur on passage through an *ODC* or *order-disorder concentration*. Additionally, preferential swelling of one phase by the solvent may skew the volume fraction of the blocks, resulting in the formation of a microdomain structure that would not be arrived at by cooling from an isotropic melt.

2.2 Influence of boundary anchoring conditions

Boundary conditions at the IMDS, which is shown schematically in Figure 3, play a crucial role in determining the structure of the mesophase in the LC BCP, and may also preclude the formation of some microdomain structures entirely. The mesogens in the LC block will typically have an orientation that they would prefer to adopt at the IMDS, if unrestrained. This anchoring condition may be homogeneous, homeotropic, or involve the mesogen lying at some angle to the IMDS. The effect of the anchoring condition is evaluated with attention paid to the size of the microdomains relative to that of the preferred mesophase, the curvature of the IMDS, the volume fraction of the mesogen in the LC block, and the strength of the anchoring in the preferred orientation. The size and form of the microdomains which would like to form determine the curvature of the IMDS. If the anchoring condition is very strong, then microdomain structures in which the satisfaction of the anchoring condition would result in a large number of LC defects may not be realized, if the strength and number density of the defects is high enough. Strong evidence for this effect is given by the reversible structural transition from spheres to cylinders on passage through $T_{I\text{-}N}$ reported by Gronski and Sanger *et al.* [23]. Alternatively, the mesophase may change, such that some other, perhaps even less ordered, mesophase is produced, but the boundary condition is satisfied and a stable microdomain structure is realized, as suggested by the data of Fischer *et al.* [11] which shows a change of mesophase from smectic A in lamellar microdomains to nematic in spherical ones. If the volume fraction of the mesogens in the LC block is small, then the preference of anchoring condition of the mesogen takes on a smaller role in determining the final structure of the LC BCP because the physical effect of the IMDS may be screened from the mesogens by the coils of their host block. If the anchoring condition is weak, then it may change to suit efficient packing of the mesogens within the microdomain into a stable mesophase.

The formation of smectic mesophases within spheres would not be expected due to the high curvature of the IMDS which excludes efficient mesogen packing in parallel layers. If the spheres were very large relative to the thickness of the smectic layers, however, the packing frustration is reduced, and smectic mesophases may indeed be observed, again subject to the volume fraction of mesogens in the LC block. The proximity of the IMDS presented by microphase

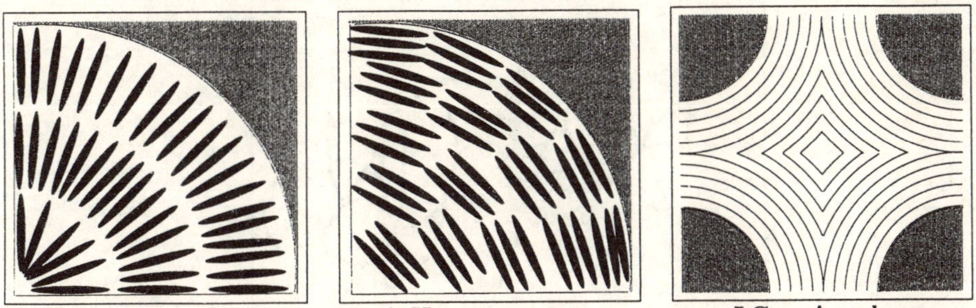

a Homeotropic b Homogeneous c LC major phase

Fig. 3 Illustration of a section of the inter-material dividing surface, IMDS, for a cylindrical microdomain in an A/B diblock copolymer. Each block occupies some area of the surface and is stretched away from it. The minority volume fraction A block is on the concave side of the IMDS

separated block copolymers makes defects in a mesophase confined by the IMDS highly unstable, depending on the anchoring condition - the degeneracy of planar anchoring at the IMDS makes it possible for defects to exist in plane within the microdomain, but not out-of-plane. Direct competition occurs between satisfying the anchoring condition and the elimination of defects due to packing which is prescribed by the IMDS anchoring, for a given mesophase structure. Possible packing frustrations for a smectic A mesophase confined to block copolymer domains are illustrated in Figure 4. The enthalpic penalty for the formationof defects in the matrix is less than that for their formation in the minority phase. Where the anchoring condition makes for efficient packing in a confining microdomain, one can expect very low levels of defects within the confined mesophase.

2.3 Predicting Morphology

The energetic penalties associated with packing mesogens into the smectic A mesophase in a restricted geometry can be assessed by considering the free energies associated with deformation of this mesophase, and the anchoring of the mesogens at the IMDS in their preferred or easy orientation. In liquid crystalline block copolymers, the confining microdomains are on the order of 10^{-1} micrometers in scale. Thus we can describe the ordering of the mesophase using Landau-de Gennes theory.

Confinement of smectic mesophases into volumes with curved surfaces

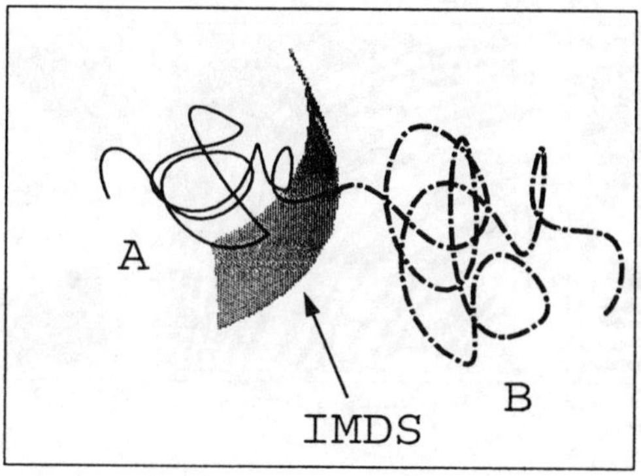

Fig.4 Packing frustration of mesogens in the smectic A mesophase confined to block copolymer microdomains. Figures are cross-sections of a quarter-sphere for mesogens in the minor ((a) and (b)) and major ((c)) domains. The mesogen length is 27 Å, mesogen spacing is 5Å, layer period is ≈30Å, the radius of curvature of the microdomains is 100 Å in each case, and the separation between the sphere centers is 500 Å. The lines shown in **c** are the layer lines for homeotropic boundary conditions and the director field for homogeneous boundary conditions. The two are related by 90°. Note that **c** is on a larger scale than **a** and **b.** Point defects of strength s=+1 exist at the centers of the spheres in **a** and **b** and a line defect of s=-1 on the diagonal between spheres in **c**

may result in the deformation of the mesophase, depending on the anchoring conditions, as mentioned above. Such spatial variation in the local order parameter of the mesophase causes elastic strains to develop within the mesophase. For smectic layers of thickness d parallel to the director **n**, there is a spatial density modulation $\rho(\mathbf{r})=\rho o\ \{1+\mathrm{Re}[\psi(\mathbf{r})\ e^{iqz}]\}$where $\psi(\mathbf{r})$ is the complex smectic order parameter and $q=2\pi/d$. The complex density order parameter, $\psi(\mathbf{r})= \eta(\mathbf{r})e^{i\phi(\mathbf{r})}$ where the positional order parameter η (**r**)describes the degree of layer ordering and ϕ (**r**) is a phase factor defining the position of the layers. Landau-de Gennes theory [24, 25] gives the free energy density written in Equation 1. The first two terms in $f(\psi,\mathbf{n})$ describe mean field contributions, i.e. they are expansions in terms of the order parameter for the free energy of the uniform mesophase that is a function of temperature only. The remaining terms depend on gradients in, and thus the spatial derivatives of the local order parameter and account for the contribution to the free energy due to deformation of the mesophase.

$$
\begin{aligned}
f(\psi,\mathbf{n}) &= \frac{1}{2}A|\psi|^2+\frac{1}{4}C|\psi|^4 \\
&\quad +c_{\|}|(\mathbf{n}.\nabla-iq)\psi|^2+c_{\perp}|(\mathbf{n}\times\nabla)\psi|^2
\end{aligned} \qquad (1)
$$

$$
\begin{aligned}
& +\frac{1}{2}\left[K_{11}(\nabla.\mathbf{n})^2+K_{22}(\mathbf{n}.\nabla\times\mathbf{n})^2+K_{33}(\mathbf{n}\times\nabla\times\mathbf{n})^2\right] \\
& +\left[-\frac{1}{2}K_{24}\nabla.[\mathbf{n}(\nabla.\mathbf{n})+\mathbf{n}\times\nabla\times\mathbf{n}]+K_{13}\nabla.[\mathbf{n}(\nabla.\mathbf{n})]\right] \\
f_e^b &= \frac{1}{2}K_{11}\left(\nabla\times\mathbf{n}\right)^2+\frac{1}{2}B\left(\frac{\delta u}{\delta z}\right)^2
\end{aligned}
$$

where A and C are coefficients of the Landau-de Gennes expansion, $c_{p//}$ is the smectic compressibility constant, related to the de Gennes compressibility B, $c_\perp$ is the bend constant associated with tilting the director away from the layer normal, and K_{11}, K_{22}, K_{33}, K_{24}, K_{13}, are the Frank nematic elasticity coefficients for director splay, twist, bend, saddle-splay and mixed splay-bend deformation. f^b_e is the free energy density associated strictly with the deformation of a *bulk* smectic A mesophase. $(\delta u/\delta z)^2$ gives the curvature of the layers of spacing u with respect to the local layer normal z. The surface energy contribution to the free energy results from deviations from ideal anchoring conditions. Thus it is a function of the difference between the actual surface order parameter and the "easy" or ideal surface order parameter. It may be explicitly expressed in the manner of Rapini and Papoular[26] in terms of anchoring angles or the corresponding vector products as shown in Equation 2.

$$
\begin{aligned}
f_S(Q^S_{ij}) &= f_S(Q^e_{ij})+\frac{1}{2}\tilde{\mathbf{W}}(Q^S_{ij}-Q^e_{ij})^2 \\
f_S &= f_S(Q^e_{ij})+\frac{1}{2}(W_\theta\cos^2\phi+W_\phi\sin^2\phi)\sin^2\theta \qquad (2) \\
F_S &= -\frac{1}{2}W_\theta\int_\Sigma(\mathbf{n}.\hat{\mathbf{z}})^2d^2\mathbf{r}+\frac{1}{2}W_\phi\int_\Sigma(\mathbf{n}.\hat{\mathbf{x}})^2d^2\mathbf{r}
\end{aligned}
$$

where Q^S_{ij} is the actual surface order parameter and Q^e_{ij} is the "easy" surface order parameter which is obtained by minimizing f_S•**z** is the easy axis in the surface plane and **x** is the direction perpendicular to the surface plane.
W_θ is the polar anchoring energy associated with out-of-plane director variations from the easy direction and W_ϕ is the azimuthal anchoring energy associated with in-plane director deviations from the easy direction, as defined in Equation 3

$$
\begin{aligned}
W_\theta &= \frac{1}{2}\left(\frac{\delta^2\gamma_s}{\delta\theta^2}\right)_\phi(\theta_s,\phi_s) \\
W_\phi &= \frac{1}{2}\left(\frac{\delta^2\gamma_s}{\delta\phi^2}\right)_\theta(\theta_s,\phi_s)
\end{aligned} \qquad (3)
$$

relative to the surface tension, γ_s. For degenerate planar anchoring, $W_f \rightarrow 0$ and $W_\phi \rightarrow \infty$.

The director field that exists within a confined mesophase is that which minimizes the excess free energy of the system, which consists of the elastic strain and anchoring energy terms, in the absence of external fields. Using the expressions for the anchoring energy density [26, 27] and the elastic strain energy provided in Equations 1 and 2, the free energy functional G[**n**(**r**)] is defined in Equation 4. Finding the equilibrium director field of the confined mesophase involves solving a variational problem that seeks the minimum of this free energy, subject to appropriate boundary conditions.

$$G[\mathbf{n}(\mathbf{r})] = \int_V f(\psi, \mathbf{n})dV + \int_\Sigma f_S(\mathbf{n})d\Sigma \qquad (4)$$

The coupling of the liquid crystal mesophase energetics to those of the block copolymer is through the size and form of the block copolymer microdomains which determine their volumes and surface areas, V and Σ. The chemical incompatibility between the backbones of the two blocks is generally enhanced by the presence of the mesogen, resulting in highly stretched chains which increase the size of the microdomains above that which one would expect from non-LC functionalized diblocks of similar molecular weight [28]. Even at temperatures above T_i, there is not a significant change in microdomain spacing which suggests that the large microdomain sizes are due more to the chemical effect of strong block segregation rather than purely a physical effect of liquid crystalline mesophase formation [19]. Any additional chain stretching that further decreases the surface to volume ratio of the mesophase and the curvature of the IMDS, thus aiding defect relaxation in the mesophase, comes with a decrease in conformational entropy of the block copolymer backbone but also a decrease in the amount of excess free energy due to the presence of the IMDS of the block copolymer. The global free energy of the system must take these contributions into account along with those for the liquid crystal mesophase. Neglecting kinetic effects, the equilibrium morphology of the system is simply that which minimizes this global free energy. Whether the anchoring condition that is observed is the thermodynamically preferred orientation that the mesogens would have at a bulk surface or is guaranteed only by the fact that the block backbones are stretched due to strong segregation, or both, is not known.

2.4 Experimental Results and Discussion

A well defined series of diblock copolymers consisting of styrene blocked with an isoprene which was side functionalized by an azobenzene based mesogen using a C_5 spacer was studied. The compositions ranged from Φ_{LC} = 0.13 to 1.0. The homopolymer LC formed a smectic A mesophase with a layer spacing of 31 Å and a packing distance of 4.4 Å. The smectic A to isotropic transition temperature was 171 °C. The compositions of the azobenzene series and the

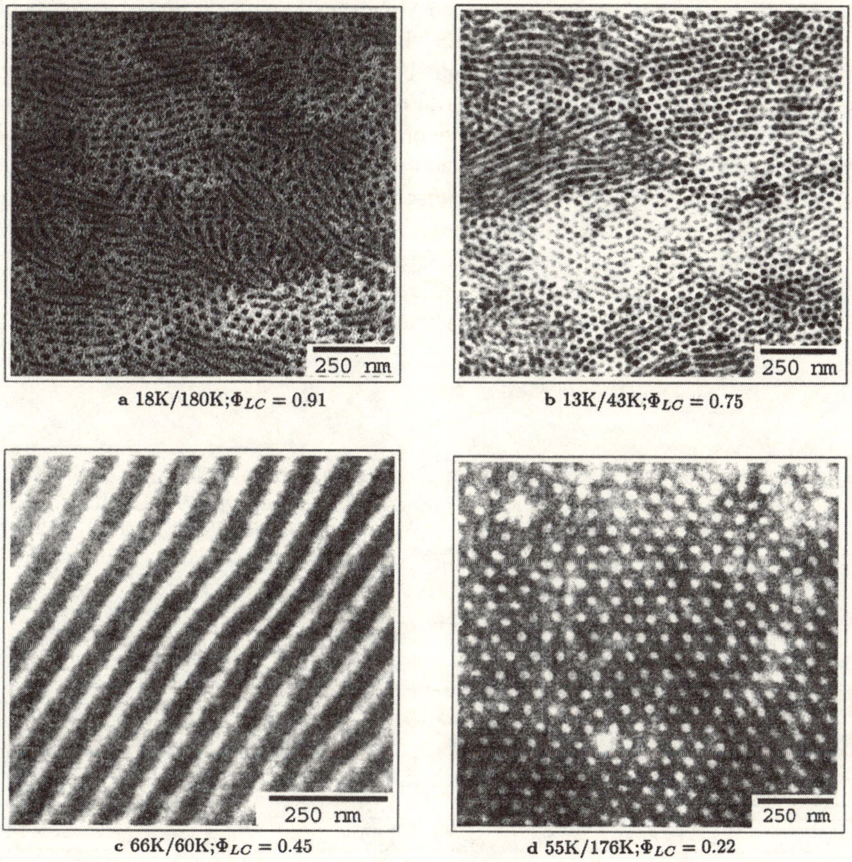

Fig. 5. Bright field TEM of bulk samples. PS regions are stained with RuO_4 and thus appear dark. **a**: PS cylinders in LC matrix; **b**: PS cylinders in L matrix; **c**: lamellae; **d**: LC cylinders in PS matrix

morphologies determined for each sample are given in Table 1 and Figure 5. Samples were prepared by slowly evaporating a 5 wt.% solution of the polymer in toluene followed by thermal annealing at 140 °C. In all cases, cylinders were found to order on hexagonal lattices as shown in Figure 5. Both small and wide angle X-ray diffraction were used to determine the orientation of the mesogens with respect to the IMDS in globally well oriented samples produced by roll-casting [29] and oscillatory shear. For cylinders of the LC block, as well as cylinders of styrene and lamellae, the mesogens were found to orient parallel to the IMDS, forming smectic A mesophases. Figure 6 illustrates the efficient packing of mesogens in smectic A layers within a cylinder. The absence of spheres of the liquid crystalline block in P(S-b-ILC) 215/36 is noteworthy. The strength of the anchoring condition of the mesogens in these materials probably excludes the formation of spheres since the packing of mesogens into the very small spheres which would form would be highly frustrated. The high volume fraction of the isoprene block constituted by the mesogens, ≈86%, further prevents the adoption of a spherical geometry. It is conceivable that the rigid mesogens may force the host block to adopt a form which has a higher surface to volume ratio than a sphere, such as a discoid, but which lowers the overall free energy of the system by satisfying the strong anchoring conditions of the mesogens at the block copolymer IMDS. Similarly, the absence of PS spheres in P(S-b-ILC) 18/180 may be due to high enthalpic penalties that would be associated with packing smectic layers around spheres.

Table 1: Summary of results obtained for various compositions of P(S-b-ILC). Φ_{PILC} indicates the volume fraction occupied by the side group LC block. M_n in kg/mol. Bulk morphologies were determined by TEM and SAXS. $T_{Sm(A)\text{-}I}$ in °C. $\Delta H_{Sm(A)\text{-}I}$ in J/g normalized by LC content

M_n [PS/PILC]	M_w/M_n	Φ_{PILC}	$T_{Sm(A)\text{-}I}$ (°C)	Morphology
215/36	1.09	0.13	118	Disordered
107/29	1.07	0.20	189	Cylinders of LC
176/55	1.13	0.22	173	Cylinders of LC
176/78	1.15	0.29	165	Cylinders of LC
66/60	1.12	0.45	153	Lamellae
8/8	1.09	0.48	135	Lamellae
59/62	1.10	0.49	158	Lamellae
27/53	1.13	0.64	165	Cylinders of PS
13/43	1.10	0.75	187	Cylinders of PS
14/72	1.24	0.82	190	Cylinders of PS
18/180	1.30	0.91	169	Cylinders of PS
78	1.15	1.00	171	Homopolymer

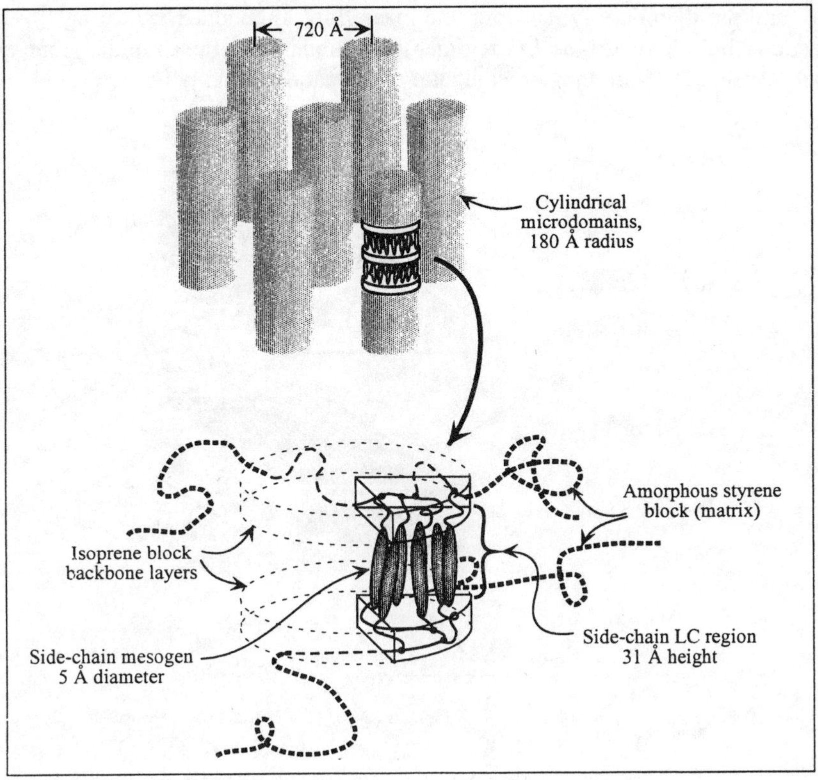

Fig. 6. Homogeneous boundary condition for smectic A mesophases confined to cylinder microdomains

The high volume fraction of the mesogens in the isoprene block of these polymers undoubtedly lends significance to the role of the mesogen's preferred IMDS anchoring condition in determining structure, as alluded to above. X-ray diffraction performed in-situ with oscillatory shear of P(S-b-ILC) 27/53 reveals this effect quite clearly. Above the clearing temperature, the poly(styrene) cylinders orient *parallel* to the shear flow. Below the clearing temperature, the cylinders lie along the neutral direction, in the *perpendicular* orientation, as illustrated in Figure 7. If the cylinders were to orient along the flow direction **x**, the lamellar-like smectic layers would then have to lie transverse to the velocity field. A more stable situation finds the smectic layers perpendicular to the velocity field, and the cylinders lying along the mass flow direction. On heating from the smectic A into the isotropic LC regime, the cylinders re-orient rapidly (time-scale about 10 minutes) from the perpendicular to the parallel orientation.

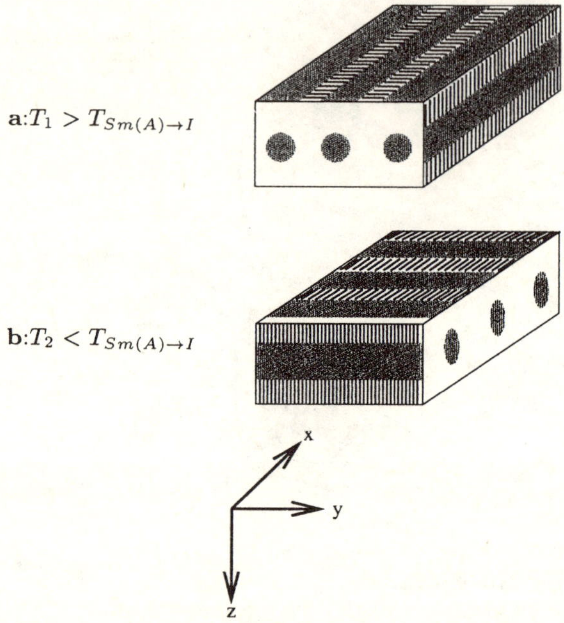

Fig. 7. Cylinder and smectic layer orientation under shear flow above and below $T_{Sm(A)\text{-}I}$ as determined by *in situ* SAXS. X-rays were incident along the neutral **z** or ∇**V** direction. Flow direction, **V** is along the **x** axis. $T_{Sm(A)\text{-}I}$ = 165 °C, T_1 = 170 °C, T_2 = 150 °C, ε = 100%, ν = 0.5 Hz.

Confinement of the mesophase to the small microdomains formed by the LC cylinders tends to eliminate defects due to the high surface to volume ratio of the microdomains, otherwise expressed in terms of the average proximity of a mesogen within the microdomain to the IMDS around it. The azimuthal spreads of the x-rays scattered by the smectic layers and by the microphase separated cylinders provide two useful estimates of the level of order within the material. The angular distribution of x-rays due to scattering from the smectic layers results from the superposition of spread due to any non-uniform *smectic layer orientation* within the microdomains as well as that due to the non-uniform *microdomain orientation*. For P(S-b-ILC) 176/55, oscillatory shear at 190 °C, ε=150%, ν= 1 Hz for 60 minutes was used to produce highly oriented cylinders within which smectic layers with high correlation of orientation between layers is observed, as demonstrated by the data in Figure 8a. An order parameter, the second component of the harmonic orientation distribution function, of S=0.93 for the cylinders and S=0.91 for the smectic layers within the cylinders is calculated in the manner of Windle [30]. In the future, micro-area electron diffraction will be applied to get structure information from single microdomains.

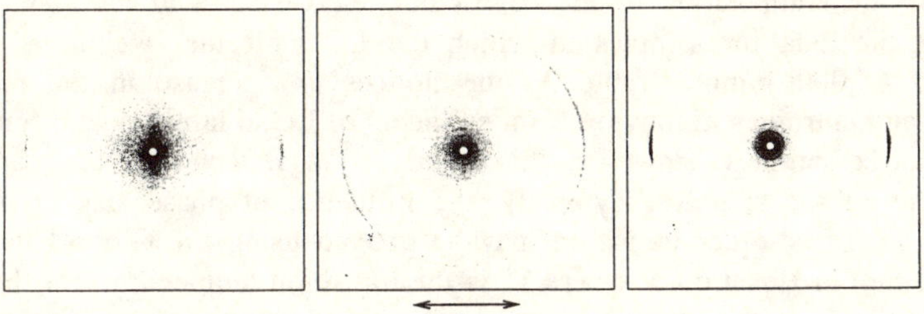

Fig. 8 SAXS of oriented samples. X-rays are incident perpendicular to the plane of shear. Flow direction indicated by arrow. (a) Highly ordered LC mesophases in cylindrical microdomains. (b) Lamellae oriented above $T_{I\text{-}LC}$ (c) Lamellae oriented below $T_{I\text{-}LC}$

The low concentration of defects present in mesophases confined to cylinders versus mesophases in the matrix or in lamellae is manifest by the optical clarity of samples in which the microstructures have been oriented either above or below the clearing temperature of the mesophase. For LC cylinders as well as PS cylinders, oscillatory shear above $T_{I\text{-}LC}$ produces films that are highly transparent at the shearing temperature. However, on cooling below $T_{I\text{-}LC}$, this transparency is lost for the LC matrix/PS cylinder sample, but retained for the LC cylinder sample. The loss of transparency is due to the formation of defects within the LC matrix or lamellae, which scatter light. In addition, in the lamellar sample, the boundary condition is degenerate - smectic layers may lie on any plane perpendicular to the surface of the lamellae and still satisfy the homogeneous anchoring condition at the IMDS. As expected, shearing below $T_{I\text{-}LC}$ produces transparent films for the matrix or cylinders of LC, and lamellae, since defects within the LC are removed by the shearing process. Heating above the clearing transition of the mesophase followed by cooling below the clearing temperature resulted in a loss of optical clarity for samples in which the LC was in the matrix or lamellae. Defects spawned readily on cooling into the LC phase, whereas repeated heating above the clearing temperature of the LC mesophases for samples in which the LC was confined to cylindrical microdomains produced no losses in optical clarity upon cooling into the LC regime. As shown in Figure 8b, an isotropic scattering ring is found at room temperature for the smectic layers in the lamellar sample which is oriented above the clearing temperature of the LC, whereas for samples oriented below $T_{I\text{-}LC}$, Figure 8c, the scattered intensity due to the smectic layers is concentrated into sharp arcs perpendicular to the lamellae, consistent with planar anchoring of the mesogens at the IMDS.

The confinement of liquid crystals is known to have effects on the thermal behavior of their systems [31]. The results for the thermal characterization of the polymers are presented graphically in Figure 9. The clearing entropy is clearly a function of the composition of the copolymer, increasing with increasing Φ_{LC}. Examining the data for samples in which the LC molecular weight is in the vicinity of 65,000 g/mol., Table 2, one notices an increase in the clearing enthalpies and entropies in moving from cylinders of LC to lamellae and finally to the LC matrix samples. However, this trend is not reflected in the transition temperatures of the samples. Typically, the influence of phase size or surface proximity on a first-order transition may be gauged using the Gibbs-Thompson relation written in Equation 5, where $T°$ is the transition temperature for the bulk material, ΔH is the heat of fusion per unit mass, σ is the mean surface free energy, V is the specific volume and l is a single parameter characterizing the smallest dimension of the phase.

$$T = T^{\circ}\left[1 - \frac{V\sigma}{l\Delta H}\right] \tag{5}$$

Table 2 Thermal data at constant LC M_n in kg/mol. $T_{Sm(A)\text{-}LC}$ in °C. $\Delta H_{Sm(A)\text{-}I}$ in J/g normalized by LC content

Sample	LC M_n	Φ_{LC}	$\Delta H_{Sm(A)\to I}$	$\Delta S_{Sm(A)\to I}$	$T_{Sm(A)\to I}$
176/55 176/78	66.5	0.27	2.20	0.50	169
59/62 66/60	61.0	0.50	3.01	0.70	156
27/53 14/72	62.5	0.74	3.75	0.83	180

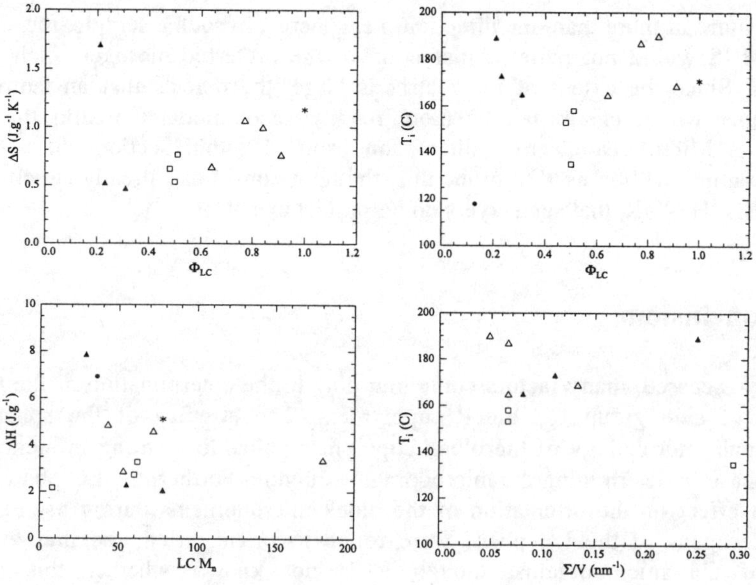

Fig. 9 Summary of DSC data for P(S-b-ILC) series. Φ_{LC}=0.13: •; PS Cylinders: Δ; PILC Cylinders: Δ; Lamellae: □; Homopolymer: *. Surface to volume ratios of the LC phase Σ/V calculated on basis of domain sizes and volume fractions.

It seems reasonable that the increase in the clearing temperature of the mesophases in cylinder forming block copolymers is due to the curvature of the IMDS, and thus lack of degeneracy in the mesogen anchoring condition, which stabilizes the mesophase by lowering the surface term, σ, in Equation 5. The influence of molecular weight can be dispensed with somewhat by considering the effect of surface to volume ratio on the clearing temperature as graphed in Figure 9. Clearly there is delineation in $T_{I\text{-}LC}$ based on IMDS curvature, i.e. between the lamellae and cylinder forming samples, as mentioned above.

It is speculated that the decrease in clearing entropy relative to the homopolymer LC is due to confinement of the mesogens to microdomains. The presence of a pseudo-isotropic layer at the IMDS may account for some of this effect, but this should not be expected given the high enthalpic penalty associated with the existence of defects within a small confined mesophase. It is more probable that such an isotropic layer may exist in samples in which the matrix is formed by the LC block, and the minority phase forms domains with surfaces of high curvature in more than one direction, i.e. spheres. In such a sample, the shape of the IMDS would not permit satisfaction of the preferred mesogen anchoring condition. Since the extent of the matrix is large, the defects that an isotropic IMDS layer would create can be more readily accommodated within the LC mesophase. Micro-area electron diffraction from ultra-thin sections of a bulk oriented sample will be used to probe this, though Gronski has already concluded, on the basis ^{2}H NMR, that such layers do not in fact exist [10].

3 Conclusion

As discussed, many factors come into play in the determination of the final structure of side group LC block copolymers. The absence of the spherical microdomain morphology of the block copolymer points to a strong influence of the LC phase in determining the microdomain structure. Further, the LC phase has a distinct effect on the orientation of the block microdomains during oscillatory shear. Mesogens of the LC phase were found to orient strictly parallel to the IMDS of the microdomains, though it is not known whether this is a thermodynamic effect, a function of chain stretching, or both. The presence of curvature in the IMDS clearly lends thermal stability to the LC mesophase whether it exists as the confined phase or as the matrix phase. In addition, confined LC mesophases were found to be of very high order due to the proximity of the IMDS to the confined mesogens, on average. The creation of functional electro-optic or piezo-electric devices using side group LC block copolymers relies on careful control of the microstructure to optimize a desired response of the material. The results of the study presented herein help to make this possible.

4 Acknowledgements

Work is supported by the NSF under grant FRG DMR 97-05271. The authors wish to acknowledge the staff of the National Synchrotron Light Source at Brookhaven National Lab including Dr. Fengji Yeh and Prof. Benjamin Hsiao, as well as Profs. Murugappan Muthukumar and Paula Hammond for helpful discussions.

References

[1] D. J. Quiram, R. A. Register, G. R. Marchand and A. J. Ryan. "Dynamics of Structure Formation and Crystallization in Asymmetric Diblock Copolymers," *Macromolecules*, 30(26):8338-8343, 1997

[2] G. Mao and C. K. Ober. "Block Copolymers Containing Liquid Crystalline Segments," *Acta Polymer*, 48:405-422, 1997

[3] T. Öge and R. Zentel. "Manipulation of Ferroelectricity in LC Polymers via Photomechanical Isomerization of Azobenzene Moieties," *Macromol. Chem. Phys.*, 197:1805-1813, 1996

[4] R. Zentel and M. Brehmer. "Create Ferroelectric Liquid Crystal Elastomers," *Chemtech*, 41-48, May 1995.

[5] G. Mao, J. Wang, C. K. Ober, M-J. O'Rourke, E. L. Thomas, M. Brehmer and R. Zentel. "Microphase Stabilized Ferroelectric Liquid Crystal (MSFLC): Bistable Switching of FLC-Coil Block Diblock Copolymers. *Chemistry of Materials*, 10(6):1538-1545, 1998

[6] W. Y. Zheng and P. T. Hammond. "Phase Behavior of New Side Chain Smectic C* Liquid Crystalline

[7] M. Brehmer, G. Mao, C. K. Ober and R. Zentel. "Ferroelectric Block Copolymers." In R. Zentel, G. Galli, and C. K. Ober, editors, *Macromolecular Symposia*, number 117, 175-179, May 1997

[8] C. Pugh and V. Percec. "Graft Copolymers Containing Thermotropic Side-Chain Liquid Crytsalline Backbones". *Polymer Preprints*, 27(1):366-367, 1986

[9] J. Adams and W. Gronski. "LC Side Chain AB-Block Copolymers with an Amorphous A-Block and a Liquid Crytsalline B-Block." *Makromol. Chem. Rapid Comm.* 10:553-557, 1989

[10] M. A. Tefehne and W. Gronski. "A Deuteron NMR Study on Order and Phase Behavior of Liquid Crystalline Copolymers." *Macromol. Rapid Comm.*, 17(5):305-311, 1996

[11] H. Fischer, S. Poser, M. Arnold and W. Frank. "On the Influence of the Morphological Structure on the Liquid Crystalline Behavior of Liquid Crystalline Side Chain Block Copolymers." *Macromolecules*, 27:7133-7138, 1994

[12] H. Fischer, S. Poser and M. Arnold. "On the Interaction of the Morphological Structure and the LC Behavior of LC Side group Block Copolymers." *Liquid Crystals*, 8(3):503-509, 1994

[13] H. Fischer, S. Poser and M. Arnold. "Liquid Crystalline Side Group Block Copolymers with n-Butyl Methacrylate as an Amorphous A-Block: Synthesis and Characterization." *Macromolecules*, 28:6957-6962, 1995

[14] G. Mao, J. Wang, S. R. Clingman, C. K. Ober, J. T. Chen and E. L. Thomas. "Molecular Design, Synthesis and Characterization of Liquid Crystal Diblock Copolymers with Azobenzene Side Groups." *Macromolecules*, 30(9):2556-2567, 1997

[15] W. Y. Zheng and P. T. Hammond. "Phase Behavior of New Side Chain Smectic C* Liquid Crystalline Block Copolymers." *Macromolecules*, 31(3):711-721, 1997

[16] P. Rangarajan, R. A. Register, L. J. Fetters, W. Bras, S. Naylor and A. J. Ryan. "Crystallization of a Weakly Segregated Polyolefin Diblock Copolymer." *Macromolecules*, 28:4932-4938, 1995

[17] A. Moment, R. Miranda and P. T. Hammond. "Synthesis of Polystyrene-Polysiloxane Side-Chain Liquid Crystalline Block Copolymers." *Macromol. Rapid Comm.*, 19(11):573-579, 1998.

[18] G. Mao. *Molecular Design, Synthesis, Characterization and Processing of Liquid Crystalline Diblock Copolymers.* Ph. D. thesis, Cornell University, 1997

[19] J. T. Chen. *Influence of an LC Block on Microstructure and Optical Properties of Block Copolymers*. Ph. D. thesis, Massachusetts Institute of Technology, 1997

[20] J. T. Chen, E. L. Thomas, C. K. Ober and S. S. Hwang. Zigzag Morphology of a poly(styrene-b-hexylisocyanate) Rod-Coil Block Copolymer. *Macromolecules*, 28:1688-1697, 1995

[21] J. T. Chen, E. L. Thomas, C. K. Ober and G. Mao. "Self-assembled Smectic Phases in Rof-Coil Block Copolymers." *Science*, 273:343-346, July 1996

[22] C. K. Ober, G. Mao, J. Wang, E. J. Kramer, J. T. Chen and E. L. Thomas. "Multiple Length Scale Self-organization in Liquid Crystalline Block Copolymers.' In R. Zentel, G. Galli and C. K. Ober, editors, *Macromolecular Symposia*, 117:141-152, May 1997

[23] J. Sanger, W. Gronski, S. Maas, B. Stuhn and B. Heck. "Structural Transition in a Nematic LC Block Copolymer Induced by the Transition to the LC Phase" *Macromolecules*, 30:6783-6787, 1997

[24] P-G. de Gennes and J. Prost. *The Physics of Liquid Crystals*. Clarendon Press, 1993

[25] A. Linhananta and D. E. Sullivan. "Phenomenological Theory of Smectic A Liquid Crystals."" *Phys. Rev. A*, 44(12):8189-8197, 1991

[26] A. Rapini and M. Papoular. "Distortion d'une Lamellae Nématique sous Champ Magnétique Conditions d'ancrage aux Parois." *J. Physique Coll.*, 30:C4-54, 1969

[27] M. Nobili and G. Durand. "Disorientation Induced Ordering at a Nematic Liquid Crystal Solid Interface" *Phys. Rev. A.*, 46(10):R6174-R6177, 1992

[28] M. Walther and H. Finkelmann. "Structure Formation of Liquid Crystalline Block Copolymers." *Progress in Polymer Science*, 21(5):951-979, 1996

[29] R. J. Albalak and E. L. Thomas. "Microiphase Separation of Block Copolymer Solutions in a Flow Field." *J. Polym. Sci. B: Polym. Phys.*, 31(1):37-46, 1993

[30] G. R. Mitchell and A. H Windle. *Developments in Crystalline Polymers*, vol. 2, chapter 3. Elsevier Applied Science Publishers, 1988

[31] G. P. Crawfors and S. _umer. *Liquid Crystals in Complex Geometries Formed by Polymer and Porous Networks*. Taylor and Francis, 1996

Crystallization Behavior and Properties of Polyolefins

R.G. Alamo, C. Chi

Florida Agricultural and Mechanical University & Florida State University College of Engineering. Department of Chemical Engineering. 2525 Pottsdamer St. Tallahassee Florida 32310-6046. USA
alamo@eng.fsu.edu

Abstract.

An overview of thermodynamic properties and crystallization kinetics of linear chains and random copolymers based on polypropylenes is given. The details of the discrete isothermal thickening of monodisperse low molecular weight species are discussed in relevance to the interpretation of the crystallization kinetics. Emphasis is given to the need to distinguish the crystallites formed from the initial melt from those transformed from them during thickening. When this distinction is made the nucleation process of either extended or folded crystallites is found to have the same temperature coefficient.

The effect of adding structural irregularities to the chain on the crystallization kinetics and lamellar morphology of high molecular weight species, is discussed by analyzing the behavior of well characterized metallocene isotactic polypropylenes. The analysis of the growth rates indicates the general behavior observed for random copolymers. However, the break found in the analysis of the kinetics according to classical nucleation theory is primarily related to a unique morphological change and may not be specific of a II - III regime transition. A dynamic morphological change due to the progressive development of the cross-hatched morphology with time has been investigated as a function of crystallization temperature, crystallization time and concentration of defects.

Introduction

Crystallization in polymers involves the development of three-dimensional order in the system and it is usually treated as a classical first order phase transition (1,2). In this review attention is focused on the role of the molecular structure and the general characteristics of the crystallite that evolve from well characterized molecules. The polyethylenes and polypropylenes are used as vehicles to reach general conclusions. These are systems for which a great deal of experimental data is available for analysis and most of the thermodynamic quantities that govern the fusion and crystallization of these systems are well known (3).

The long chain character of polymeric systems carries the difficulty in extracting ordered sequences of sufficient length from the highly entangled disordered melt to insure thermodynamic stability of the evolving crystallite. In contrast, the crystallization of a hydrocarbon or low molecular weight monodisperse specie occurs very rapidly at temperatures just below their melting temperature. The study of long chain n-alkanes as model compounds of regular chain length allows for further understanding of the thermodynamic processes in general and of polymers in particular. A direct companion can be made of the crystallization and melting processes of n-alkanes and the more polydisperse molecular weight polyethylenes of the same length. The relevance and importance of these studies is manifested in the ongoing structural studies of long chain n-alkanes (4,5). Once the details of the thermodynamic behavior of these two systems are well developed, one could extrapolate the behavior for understanding the crystallization and melting of the more polydisperse, commercial type, high molecular weight polymers.

Extension of the studies to high molecular weight linear species have allowed the evaluation of the strong effect of the length of the chain, chain entanglements and polydispersity on the crystallization behavior of linear polyethylene fractions (6-8) and samples with narrow molar mass distribution synthesized with metallocene-type catalysts (7). Understanding the properties of low and high molecular weights species as model compounds is important to establish a framework that can be used to study the effect of the addition of co-monomeric units into a chain. Within this context, metallocenes isotactic poly(propylenes), iPP, offer the possibility to study the effect of chain length and type, content and distribution of defects inserted in the chain on the properties of these materials. These studies were hampered in the Ziegler type iPP from the strong correlations existing between the mentioned structural variables.

Low Molecular Weight Species. ***N-Alkanes***

The comparison of structural and thermodynamic properties of n-alkanes and low molecular weight polyethylenes, in the most important structural form for the analysis of semicrystalline polymers: the folded structure, was made possible through the synthesis of high molecular weight n-alkanes (10,11). N-alkanes with carbon numbers ranging from 102 to 390 provided the set of model compounds to study the crystallization behavior of chain molecules.

Due to the different synthetic path, the heterogeneity in chain lengths of the polymeric-type of chain forbids pairing of the ends of extended crystals conferring an inherent disordered surface to these crystals. This contrasts with a more regular surface of the n-alkanes crystals where the ends are paired. In spite of these major differences in the crystallite structure, it was found that the melting temperatures of extended n-alkanes and low molecular weight PE crystals of the same length are the same (12). The fact that they both present the same melting temperature is correlated with a disorder of the ends of the chains just before they melt in both types of crystallites (12). Thus, an important first connection between extended n-alkanes model compounds and polymers was found in these results. A

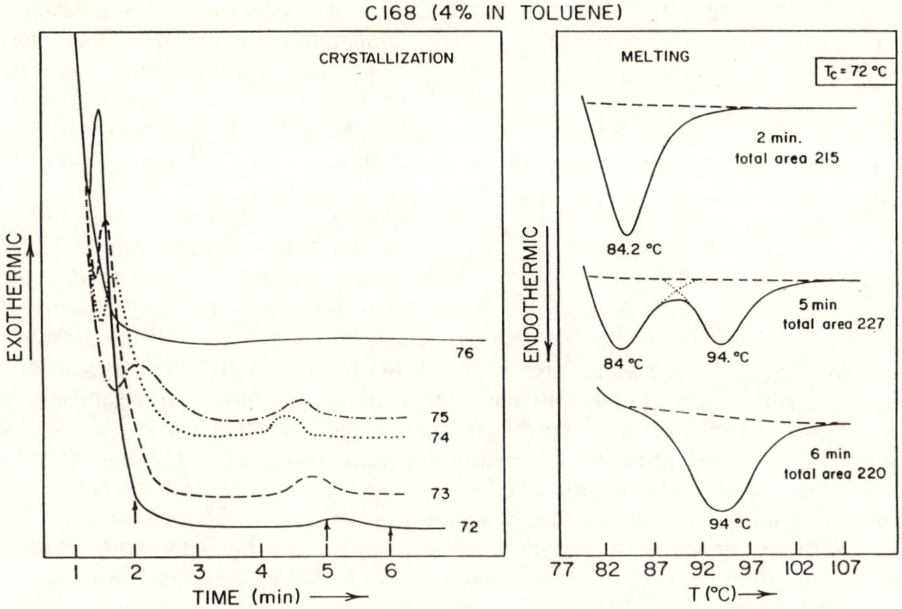

Fig. 1. (a) DSC exotherms for indicated Tc. (b) DSC endotherms at indicated Tc and times (From Reference 13)

second connection is the very similar molecular weight at which they both start to fold (~ 2000 g/mol) when crystallized from the melt and from solution (12).

The study of the crystallization and morphology of n-alkanes with carbon numbers above 150 revealed an unprecedented feature, a discrete isothermal thickening along the crystallization process (13 - 15). An example is given in Fig. 1 for the crystallization of C_{168} H_{338} from a 4% solution in toluene. Two distinct exotherms are present in a narrow crystallization temperature interval (Fig. 1a). The exotherm that appears at the lowest crystallization times follows the expected kinetic behavior, with a retardation in the rate for increasing temperatures. The behavior of the second exotherm is peculiar for it is shifting at lower times with increasing temperatures. Rapid melting of the crystals formed at different times during crystallization (without cooling) is crucial to understanding the cooperative isothermal thickening. An example is shown in Fig. 1b for crystallization at 72°C. The crystallization at this temperature is quite fast and most of the transformation has occurred after 2 minutes of crystallization. The dissolution temperature of the crystallites formed after 2 min. is in agreement with the theoretically expected for once folded crystals (15). With increasing crystallization time, the second endotherm appearing at higher temperatures indicates that an isothermal transformation from the once folded crystals to extended crystals is occurring within a conserved system (the total heat of fusion remains constant before and after thickening). This transformation is predicted to occur through a cooperative melting- recrystallization process. The recrystallization is indicated by the presence on the second exotherm in Fig. 1 a. The kinetically favored folded structures melt and recrystallize into thermodynamically more stable extended crystals. It is also observed that the rate of transformation increases exponentially

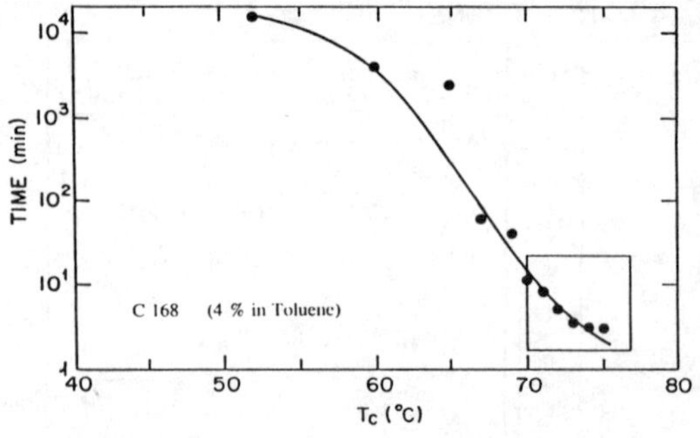

Fig. 2. Time for complete transformation from folded to extended crystals

with increasing crystallization temperature and, therefore, has important implications in the analysis of the crystallization kinetics. Figure 2 shows the time required to obtain complete transformation from the folded to extended crystals as a function of crystallization temperature for the same n-alkane. It is clear that the transformation is very rapid at crystallization temperatures above 72°C. At the highest temperatures the formation of folded crystals and their transformation is simultaneous. It is thus, important to establish the temperature range in which the rate of thickening is comparable to the rate of formation of folded structures by performing simultaneous DSC melting analysis. This allows for the identification of a temperature region where the crystallization kinetics of pure structures can not be quantified (16). In a different work (4, 17) the exotherm appearing at longer times in Fig. 1a has been identified with a poisoning effect by folded crystals into a growing extended surface; however, in the temperature range where the crystallization kinetics were studied, the thickening behavior was not analyzed.

Analysis of the crystallization kinetics of long-chain n-alkanes according to classical nucleation theory (16), indicate that the nucleation process of pure folded and pure extended crystallites is the same and shows similar characteristics to those found in low molecular weight polyethylenes.

High Molecular Weight Species. *Metallocene iPPs*

The influence of molecular weight on the crystallization kinetics and morphology of molecular weight fractions of linear polyethylene has been studied in detail (6,7,18). It is now well documented that increasing molecular weight retards the rate of crystallization and lowers the extent of transformation, i.e., the degree of crystallinity that can be attained at a given crystallization temperature due to the increased number of entanglements in the melt. The change in melt topology with changing molecular weight is also reflected in different lamellar crystallites arrangements. Thin rods from individual lamellae are observed for the lowest molecular weights (Mw~1200 g/mol), while for Mw up to 100,000 g/mol well formed spherulitic structures are formed. The lamellae formed from the highest molecular weights did not show any specific organization (18). Similar changes in the overall scattering patterns from different lamellar arrangements with molecular weight have also been observed with poly(ethylene oxide) and isotactic polypropylene (19, 20).

Attention has been focused recently on the effect that structural irregularities introduced in the chain have on the properties of these copolymeric type systems. The molecule of isotactic polypropylene can be considered a copolymeric type. The stereo or regio type of defets inserted randomly in the chain

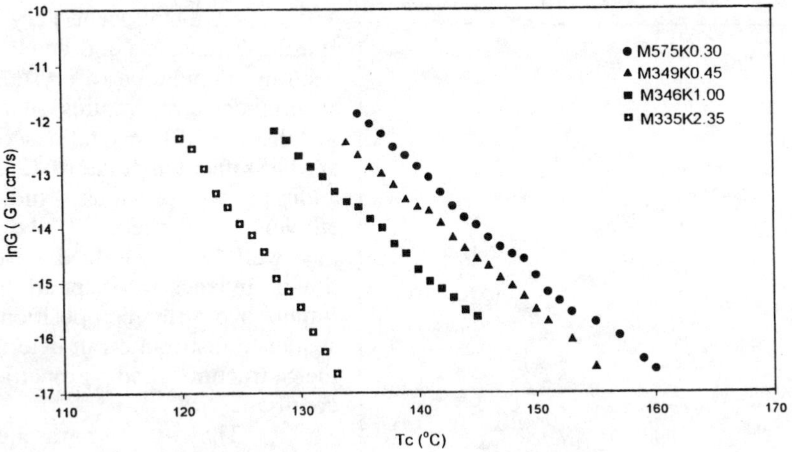

Fig. 3. Variation of growth rate with crystallization temperature for metallocene polypropylenes with different concentration of defects.

can now be controlled using metallocene catalysts in the synthesis. Thus, the properties of isotactic polypropylenes can be studied as a function of the type and the fractional content of these defects for a fixed molecular weight and molecular weight distribution. These homogeneous metallocene catalysts produce a more uniform defect distribution in the iPP with a narrower molecular weight distribution than the iPP synthesized with classical Ziegler-type catalysts. The kinetics of the spherulitic growth and morphological changes can be studied as a function of a specific chain microstructure. Fig. 3 shows spherulitic linear growth rates as a function of crystallization temperature for a series of metallocenes iPP with defect content increasing from 0.3 to 2.35 mol%. Generally for the crystallization of polymers and in agreement with a nucleation controlled process, the growth rates of all iPP's show a strong negative temperature coefficient with values that decrease about 100 times when the crystallization temperature increases from 135°C to 160°C. Although the crystallization temperature interval of the four samples shown in Figure 3 do not overlap, however, a generalization could be made for the variation of the growth rate of these metallocenes. The growth rate decreases about two orders of magnitude when the total defect content is increased from 0.3 to 2.35 mol%. Thus, typical random type copolymer behavior is observed with the defects in the polypropylene molecule acting as mainly non-crystallizable units. An increase in non-crystallizable structural irregularities leads to a reduction of the amount of crystallizable sequences and, thus, to a decrease in the feasibility and rate of crystallization (21, 22). The behavior is analogous to the overall crystallization rate of a series of random ethylene copolymers (22). For these copolymers the rate was also found to decrease about 2 orders of magnitude with increasing concentration of branch points from 1.21 to 2.3 mol%.

Figure 4 shows the growth rates of a metallocene iPP and a fraction from a Ziegler iPP with very similar total defect content (~0.5 mol%). At any crystallization temperature the growth rate of the Ziegler fraction is systematically higher than that of the metallocene reflecting that the Ziegler fraction retains an inter-chain distribution that is broader than that of the metallocene. The fractional content of almost defect free molecules of the Ziegler fraction (which are absent in

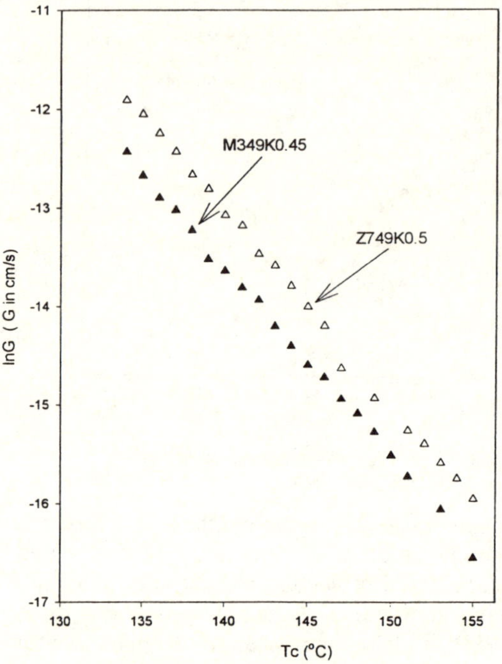

Fig. 4. Growth rates versus Tc of a metallocene iPP and a Ziegler fraction with similar content of defects.

the metallocene) crystallize at a faster rate.As a consequence, the resulting population of crystals melt at higher temperatures than the metallocene iPP crystallites formed at the same temperature (23). In addition to previous studies in ethylene copolymers (22), the results in well characterized isotactic polypropylenes support the importance of the compositional and sequence distribution in determining the structure and properties of copolymers.

The intraspherulitic lamellar arrangement was also studied as a function of structural irregularities in the poly(propylene) molecule. The more polydisperse Ziegler type show, for crystallization temperatures lower than 134°C, a morphology of spherulites with positive birefringence consisting of dentritic arrays of intercrossing lamellae (cross-hatching morphology). Spherulites with negative birefringence were obtained in the high temperature region and associated with preferentially radiating lamellae (24).

A detailed study of the lamellar morphology with the series of metallocene iPP shown in Fig. 3 led to additional and unique results (21). Figure 5 is an example of the evolution with time of the spherulitic morphology of sample M575K0.3 at a fixed crystallization temperature. The lamellar arrangement during crystallization is such that a dynamic change in the birefringence is observed with time. At temperatures higher than 145°C negative spherulites develop at the early stages of the crystallization, however, with prolonged crystallization the spherulites become mixed and occasionally positive. The change in birefringence, related to a variation of the relative fractions of radial and tangential lamellae, indicates clearly that even at crystallization temperatures at and above 160°C, a large population of tangential lamellae has grown only after prolonged crystallization. If the crystallization is not carried out for long times at the highest crystallization temperatures, predominantly radial lamellae will be observed (25, 26). An interesting feature is the conserved linearity of the variation of the radius of the spherulite with time throughout the total transformation (even after impingement). As seen in the example of Fig. 5, a relatively sharp transition in the morphology of the spherulites at the early stages of the crystallization is found for all the iPPs with defects content up to 1.5 mol%.

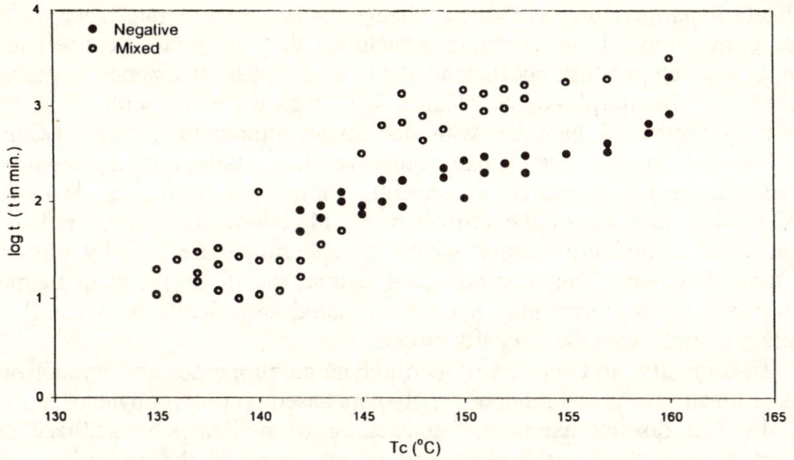

Fig. 5. Overall birefringence change as a function of crystallization temperature and crystallization time for M575K0.3.

The early formed spherulites change from mixed to negative at a temperature which decreased with increasing defect concentration. This temperature is listed in Table I as T_{change}. Axialites with positive birefringence were obtained for the iPP with 2.35 mol% of defects in the whole temperature range.

Table I. Regime and morphological transitions

IPP	T_{change} (°C)	T_{break} (°C)
M575K0.30	145	146
M349K0.45	140	145
M346K1.00	139	140
M335K2.35	---	---
Z651K0.3	145	145

Also listed in this table is the temperature at which a break consistent with a transition from Regime III to Regime II in the crystallization kinetics is obtained (21). The kinetics were analyzed using classical coherent surface nucleation theory applied to the experimental growth rate data. The typical data for U* (1500 cal/mol), $T\infty$ (Tg-30°) and T_m° (200°C) were used in the analysis (21, 27). It is obvious that the temperature at which a morphological change is obtained at the early stages of the crystallization, is the same as the temperature of the break resulting when nucleation theory is applied to the kinetic data.

No break was found in sample M335K2.35. The correlation between growth regime and morphology has been associated with different degrees of adjacent reentry (27). A high degree of tight folding was suggested to be present in the lamellae of negative spherulites growing in regimes I and II and a connection was suggested between the formation of transversal lamellae and the enhancement of non-adjacent reentry in regime III. However, the fact that we observed a change from negative to mixed and positive spherulites with crystallization time, at a fixed crystallization temperature, without a change in the growth rate, rules out the previous correlation. It is, therefore, concluded that the break obtained in the analysis of the temperature coefficient of the crystallization kinetics is mainly a consequence of the morphological change. The break correlates with an enhanced formation of tangential lamellae with decreasing temperature, from the initial stages of crystallization. The change in the overall crystallographic growth plane when more profuse development of tangential lamellae is occurring , leads to an increase in the variation of the growth rate with temperature compared to the variation in the temperature region where the spherulites are initially formed of mainly radial lamellae. This morphological feature, specific to the unique lamellar branching in polypropylene may not be correlated specifically with the II - III growth regime transition theoretically proposed.

In summary, an overview of thermodynamic properties and crystallization kinetics of linear chains and random copolymers based on polypropylenes has been presented The discrete isothermal thickening of n-alkanes crystallized from solution and from the melt is discussed in relevance to the analysis of their crystallization kinetics. The isotactic polypropylenes properly treated as random copolymers from the crystallization point of view, were discussed in the study of the effect of structural irregularities on the spherulitic growth and lamellar morphology of these copolymers. The general characteristics of the kinetics and unique features of the morphology of isotactic polypropylenes with varying concentration of defects are discussed.

Acknowledgement

This work was supported in part by the National Science Foundation, Polymers Program. Grant DMR - 94-19508 and by the EXXON Education Foundation.

References.

1. P.J. Flory, *J. Chem. Phys.*, **17**, 223, 1949.
2. L. Mandelkern, *Crystallization of Polymers*, McGraw Hill: New York, 1964, pp. 20-37.
3. Mandelkern L., R.G. Alamo. "Thermodynamic Quantities Governing Melting". in *"American Institute of Physics Handbook of Polymer Properties"*, J. Mark Ed.; AIP Press: New York, 1996.
4. S.J. Organ, P.J. Barham, M.J. Hill, A. Keller, R.L. Morgan, *J. Polym. Sci., Polym. Phys. Ed.*, **35**, 1775, 1997.
5. D.C. Basset, R.H. Olley, S.J. Sutton, A.S. Vaughan, *Polymer*, **37**, 4993, 1996.
6. E. Ergoz, J.G. Fatou, L. Mandelkern, *Macromolecules*, **5**, 147, 1972.
7. J.D. Hoffman, L.J. Frolen, G.S. Ross, J.I. Lauritzen, *J. Res. Nat. Bur. Stands.*, **79A**, 671, 1975.
8. J.G. Fatou, C. Marco, L. Mandelkern, *Polymer*, **31**, 1685, 1990.
9. J.G. Fatou, R. G. Alamo, L. Mandelkern (to be published).
10. I. Bidd, D.W. Holdup, M.C. Whiting, *J. Chem. Soc., Perkin Trans.* **1**, 2455, 1987.

11. K.S. Lee, G. Wegner, *Makromol. Chem., Rapid Commun.* **6**, 203, 1985.
12. L. Mandelkern, A. Prassad, R.G. Alamo, G.M. Stack, *Macromolecules,* **23**, 3696, 1990.
13. Alamo, R.G. "The Crystallization Behavior of Long Chain N-Alkanes and Low Molecular Weight Polyethylenes" in *NATO ASI-C Series, 'Mathematical and Physical Sciences Vol. 405: Crystallization of Polymers'.* M. Dosiere Ed. Klumer Academic Pub, Boston 1993.
14. G. Ungar, S.J. Organ, *J. Polym. Sci., Polym. Phys. Ed.,* **28**, 2353, 1990.
15. Alamo R.G., L. Mandelkern, G.M. Stack, C. Kronke, G. Wegner. *Macromolecules* **26**, 2743, 1993.
16. Alamo R.G., L. Mandelkern, G.M. Stack, C. Kronke, G. Wegner. *Macromolecules*, **27**, 147, 1994.
17. G. Ungar, A. Keller, *Polymer*, **28**, 1899, 1987.
18. J. Maxfield, L. Mandelkern, *Macromolecules*, **10**, 1141, 1977.
19. R.C. Allen, L. Mandelkern, *J. Polym. Sci., Polym. Phys. Ed.*, **20**, 1465, 1982.
20. R.C. Allen, L. Mandelkern, *Polym. Bull.* **17,** 473, 1987.
21. Chengyou Chi, Thesis. Florida State University, 1997.
22. Alamo R.G., L. Mandelkern. *Macrolecules*, **24**, 6480, 1991.
23. Isasi, Kim, Alamo, Mandelkern (to be published).
24. F. L. Binsbergen, B.G.M. De Lange, *Polymer,***9**, 23, 1968.
25. K. Yamada, S. Matsumoto, K. Tagashira, M. Hikosaka, *Polymer*, **39**, 5327, 1998.
26. D.C. Bassett, R.H. Olley, *Polymer*, **25**, 935, 1984.
27. B. Monasse, J.M. Haudin, *Colloid and Polym. Sci.* **263**, 822, 1985.

Modeling of Side-Chain Liquid Crystal Polysiloxanes

B. L. Farmer* and G. M.Podojil+

Department of Materials Science and Engineering, University of Virginia, Charlottesville, VA 22903-2442 USA farmerbl@ml.wpafb.af.mil

R. Pachter and W. W. Adams

Materials and Manufacturing Directorate, Air Force Research Laboratory, Wright-Patterson Air Force Base, OH 45433 USA pachterr@ ml.wpafb.af.mil

Abstract

Molecular dynamics (MD) simulations of polysiloxanes having cholesterol and methoxynitrostilbene (MONS) substituents tethered to the siloxane backbone by an allyloxybenzoate spacer group demonstrated that ordering involving all of the substituents developed quite rapidly, even though only the cholesterol group is by itself inherently mesogenic. The observed ordering varied with the nature and arrangement of the substituents. Generally, strong interactions between cholesterol groups resulted in the longest-lived pair-wise associations. MONS substituents also showed orientation, arising primarily from their associations with cholesterol mesogens. The overall order parameter was typically unaffected by the incorporation of MONS at fractions of the order of 25%. The polymer having alternating MONS and cholesterol substituents did not develop any significant order. Association lifetimes were not simply related to order parameter. Polymers which achieved the highest order parameters did not exhibit the longest association lifetimes. Cholesterol - cholesterol associations tended to be the longest, MONS - MONS associations the shortest.

Introduction

Side chain liquid crystalline polymers (LCPs) may provide a means both to optimize and to stabilize the orientation of substituents having useful optical properties. Above the glass transition of the polymer, the liquid crystalline character of the material can induce a preferred orientation of a chromophore. That orientation can then be preserved by reducing the temperature to bring the material into its glassy state. Notably, even low molecular weight cyclic siloxanes having

*Current address: Air Force research Laboratory, AFRL/MLBP, 2941 P St., Suite 1, Wright Patterson AFB, OH 45433 USA
+Current address: Kent Displays, Inc., Kent, OH 44240 USA

cholesterol and biphenyl substituents demonstrate this temporal stability of induced ordering [1,2]. Previous molecular dynamics simulations examined the behavior of cyclic and polymeric siloxanes having cholesterol and biphenyl substituents [3,4]. Results indicated that mesogen ordering occurred quite rapidly and the order parameter was comparable to that observed experimentally for related materials. Characterizing the influence that a chromophore {specifically, methoxy-nitrostilbene (MONS)} has on the LC properties and the ability of the LC to induce orientation of the chromophore are the objectives of this study.

Methods

Molecular dynamics (MD) simulations were carried out using SYBYL [5] and an augmented and modified version of the Tripos force field [6]. Specifically, parameters for the siloxane backbone were introduced and modifications were made to the torsional potentials to give better representation of the conformational behavior of biphenyl and MONS. These changes in the force field have been described previously [4]. For consistency with earlier simulations, Coulombic interactions were not included.

MD simulations of single polysiloxane molecules containing 15 or 16 repeat units were carried out. Various combinations of cholesterol {C}, biphenyl {B}, and MONS {M} moieties were attached to the polymer backbone with an intervening allyloxybenzoate spacer. Molecules having all {C}, all {M}, alternating {C} and {M}, a repeating sequence of {M}/4{C}, and a repeating sequence of {C}/{B}/{C}/{M} were examined. (These are designated 16C, 16M, M/C, M/4C, and C/B/C/M hereafter). The generic structures for the polymer and substituents are shown in Fig. 1. The results from other combinations have also been reported. [4]

Simulations began with a rapid increase in temperature from 0 to 300 K. The system was allowed to equilibrate for 150 ps before the data acquisition period of 50 ps. Time steps of 1 fs were used, with coordinate data being saved at intervals of 5 ps for subsequent analysis. A direction vector was defined for each substituent (excluding the spacer) and the overall order parameter was computed for the system at these time intervals. The order parameter is given by

$$S = \tfrac{1}{2}(3\langle\cos^2\alpha_i\rangle-1) \tag{1}$$

where α_i is the angle between the substituent direction vector and the overall director (i.e. the average of the individual direction vectors). A value of 1 denotes perfect alignment, 0 denotes a random arrangement, and –1/2 denotes a random arrangement in a plane perpendicular to the director.

The order parameter yields a good picture of the overall behavior of the system, but information on individual pair associations is lost. To characterize these associations, normalized vector dot products were calculated for all possible substituent pairings at 5 ps time intervals, yielding a level of alignment for each pair as a function of time. An arbitrary minimum dot product value of 0.75 was

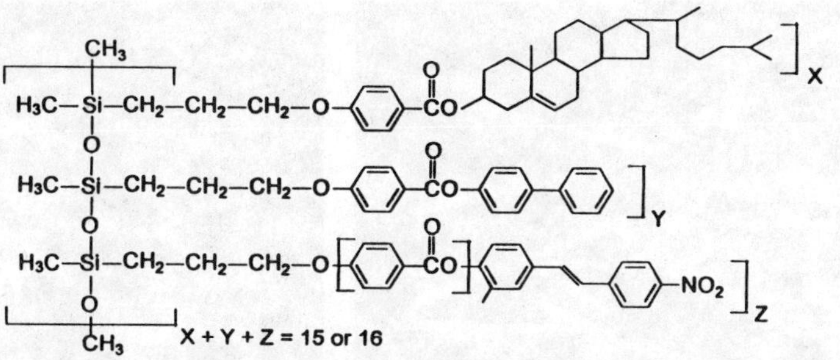

Fig. 1. General form of the substituted polysiloxanes: X, cholesterol; Y, biphenyl; Z, MONS

used as the criterion for a pair to be considered aligned. This corresponds to being aligned within about 40 degrees. This rather wide range ensures that brief thermal angular fluctuations do not affect the tabulation of lifetime data. The lifetimes of interactions were tabulated, and averages were determined for the different interactions (e. g. C-C, C-M, M-M etc.) in the various molecules. Only pairs that remained aligned for a minimum of 10 ps were counted. The second moments of the distributions of alignment lifetimes were used to compute the averages since the longer-lived interactions are of most interest.

As previously [4], a simplified view of a molecule is used for visualization. Each substituent is replaced by a bar representing its direction vector. Similarly, the backbone and spacers are simplified by reducing them to a collection of simple linear segments. For molecular systems of 1000 to 1500 atoms, this provides a clear depiction of mesogen alignment. It is important to note that this is merely a viewing mode applied after simulations had completed. Each MD simulation explicitly considered all atoms in the molecule.

Results and Discussion

Results on other substituent arrangements, reported previously[4], are needed for comparison with the results on MONS-substituted polysiloxanes reported here. For the all cholesterol-substituted molecule (16C), the order parameter was 0.29 and the average association lifetime was 117 ps. The polymer having alternating biphenyl and cholesterol substituents (B/C) gave an overall order parameter of 0.41, 0.47 for the biphenyl substituents alone, and 0.36 for the cholesterol substituents alone. Association lifetimes were: overall, 78 ps; {B}-{B}, 55 ps; {C}-{C}, 66 ps; and {B}-{C}, 91 ps.

All MONS Molecule

The 16M molecule showed significant alignment of the MONS substituents. Figure 2 shows orthogonal rod depictions of this molecule after 200 ps of MD simulation. Most of the substituents aligned on one side of the siloxane backbone, taking advantage of the flexible allyloxybenzoate spacer group. The order

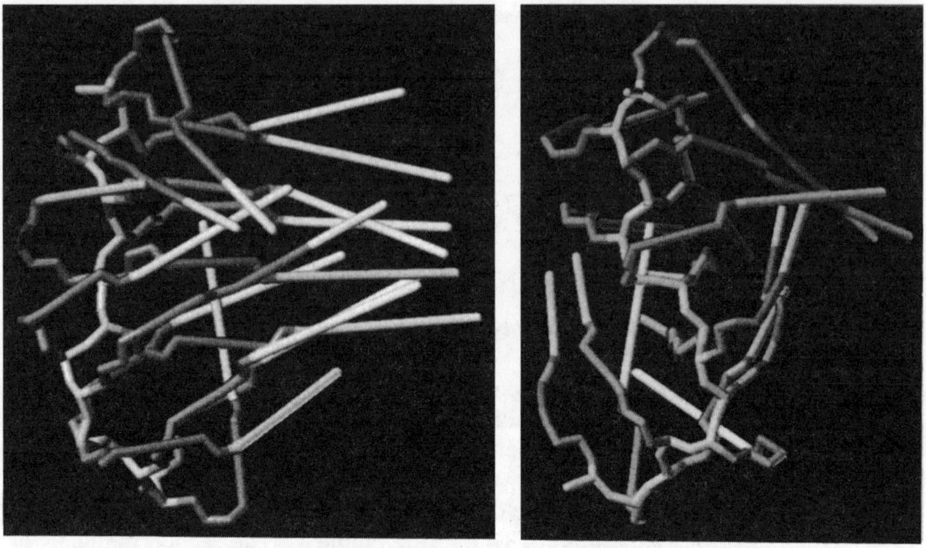

Fig. 2. Orthogonal views of the 16M molecule after 200 ps of MD simulation.

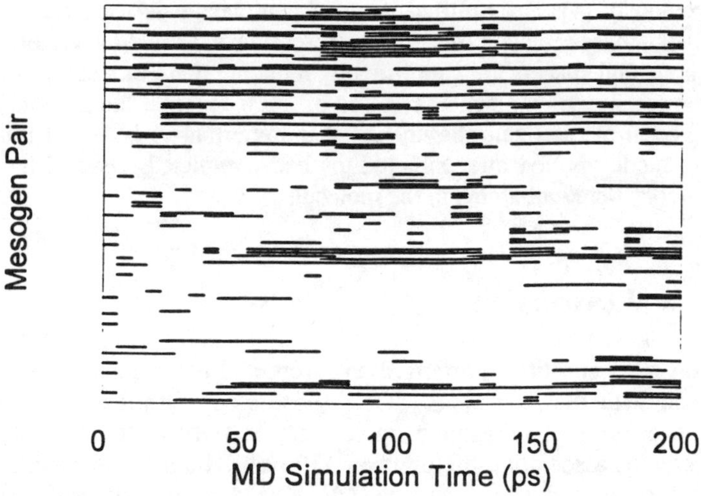

Fig. 3. Mesogen associations during the MD simulation for the 16M molecule.

parameter for MONS varied between 0.23 and 0.51 with an average value of 0.36, indicating a moderate level of order for the substituents. Figure 3 shows mesogen associations for the 16M molecule as a function of MD simulation time. In this plot, each horizontal bar represents a mesogen pair associated over the time interval indicated. The weighted average lifetime of associated {M}-{M} pairs was 57 ps, which is lower than that observed previously for {B}- {B} and {C}-{C} pairs in single mesogen type molecules (68 ps and 117 ps, respectively) [4], indicating a somewhat more fluid association.

Alternating {M}/{C} Molecule

The M/C molecule displayed poor ordering behavior, with the order parameter for cholesterol ranging from -0. 12 to 0.25, averaging 0.07, and the value for MONS ranging from -0.04 to 0. 10 and averaging 0.02. When all of the mesogens are considered together, the average order parameter was 0.05. These averages indicate no stable overall order. This is different from the alternating B/C molecule studied previously, where both biphenyl and cholesterol displayed moderate levels of order. It is interesting that both cholesterol- and MONS-substituted homopolymers had moderate levels of ordering by themselves, but when these substituents were mixed in an alternating fashion, the ordering behavior was no longer observed.

Tracking the pair associations in this molecule revealed that the fraction of pairs that were associated was small, but that their lifetimes were fairly long. The average values of lifetimes were 68 ps, 77 ps, and 81 ps for the {C}-{C}, {M}-{M}, and {C }- {M} pairs, respectively. The overall average was 77 ps.

Repeating {M}/4{C} Molecule

The M/4C molecule developed fairly large stable clusters of cholesterol mesogens. The MONS substituents appeared to be associated with the cholesterol to a reasonable degree, but did not appear to influence the alignment behavior of the cholesterol. The backbone was observed to take on a folded conformation to accommodate the mesogen alignment.

The order parameter for cholesterol ranged from 0.21 to 0.49 with an average of 0.31 while the value for MONS ranged between 0.19 and 0.42 with an average of 0.28. These values are of the order observed for the respective single mesogen molecules. Perhaps more importantly, the order parameter for all mesogens considered together is 0.28; roughly the same as those for MONS and cholesterol each considered separately, indicating that the MONS and cholesterol are aligned in a cooperative fashion.

The average lifetimes for the {C}-{C}, {M}-{C}, and {M}-{M} pair types were 79 ps, 56 ps, and 5 ps, respectively. The value for the {M}-{C} pair is lower than that observed in the M/C molecule. The extremely low value for the {M}-{M} pair is not surprising, given that there are only 3 substituents being considered. These substituents are not actively associating with each other, but merely follow along with the cholesterol alignment. The average lifetime for all pair types, 72 ps, is comparable to that for the other mixed mesogen molecules.

Repeating C/B/C/M Molecule

For the C/B/C/M molecule, all but two of the mesogens associated into a single large cluster oriented perpendicular to the siloxane backbone. Biphenyl, cholesterol, and MONS substituents all intermingled. This mixture of substituents was conducive to cooperative behavior and coalescence into a large-scale ordered structure.

The order parameter data reflect the ordering observed. (Note that the order parameter is for the two smaller substituents taken together.) In the C/B/C/M molecule, the order parameter for biphenyl and MONS ranged from 0.26 to 0.45

and averaged 0.35 while that for cholesterol ranged from 0.33 to 0.46 and averaged 0.37. The order parameter for all mesogens considered together averaged 0.36. These values indicate moderate levels of order for all of the mesogens. In addition, the closeness of the three values reflects the intermingling and cooperative alignment of the different mesogens.

The average lifetime for {C}-{C} pairs was 149 ps, higher than in any other molecule studied, indicating that the mixture with biphenyl and MONS served to stabilize the {C}-{C} interaction in this molecule. {B}-{B} and {M}-{M} lifetimes were 9 ps and 109 ps, respectively. The average lifetimes for mixed substituent pairs, {B}-{M} and {C}- {M}, were 111 ps and 108 ps, respectively, indicate stable mixed mesogen pairs as well. The average for all pair types (105 ps) was higher than in any other molecule

Conclusions

These simulations suggest that the incorporation of a chromophore such as MONS at concentrations of the order of 25% do not disrupt the organization of side-chain cholesterol mesogens in these polysiloxanes, and in fact may provide fluidity which allows higher levels of order to develop. It was found that high levels of order and pairwise stability of substituents do not necessarily go hand in hand. Individual associations within an ordered cluster may be quite fluid in nature and tightly formed associations may hinder the development of larger scale order. Associations between pairs of cholesterol substituents dominate the structures which develop. The associations of MONS substituents with cholesterol substituents tend to be longer lived than those between pairs of MONS substituents. A molecule incorporating cholesterol, biphenyl and MONS substituents onto the same polysiloxane gave the best organized structure observed in the simulations.

References

1. T. J. Bunning, H. E. Klei, J. T. Koberstein, E. T. Samulski, W. W. Adams, and R. L. Crane, USAF Technical Report No. WL-TR-92-4051, 1992
2. S. G. McNamee, T. J. Bunning, C. M. McHugh, C. K. Ober, and W. W. Adams, *Liq. Cryst.* **17**, 179 (1994)
3. R. Pachter, T. J. Bunning, R. L. Crane, W. W. Adams, E. P. Socci and B. L. Farmer, Makromol. Chem., Theory Simul., 2, 337 (1993).
4. G. M. Podojil, B. L. Farmer, and W.W. Adams, *Polymer,* **37(10)**, 1825 (1996)
5. SYBYL, Version 6.01, Tripos Associates, Inc., St. Louis, MO, 1992
6. M. Clark, R. D. Cramer III, and N. Van Opdenbosch, *J. Comp. Chem.*, **10**, 982 (1989)

Polymorphic and Polytypic Transformations during Crystallization of Long-Chain Compounds

Fumitoshi Kaneko

Department of Macromolecular Science, Graduate School of Science, Osaka University, Toyonaka, Osaka 560-0043, Japan
toshi@chem.sci.osaka-u.ac.jp

Abstract. Crystallization processes of long-chain compounds have been studied with FT-IR spectroscopy, which is highly sensitive to structural differences in functional groups. It has been found that several types of structural changes take place during the crystallization of fatty acids. In the solution crystallization of stearic acid, single crystals of the E form transform to the B form. This transformation starts with a heterogeneous nucleation of the B form on the {0 0 1} faces of the single crystal of E, and the B domains spread to the whole region of the single crystal through a solid-state phase transition. Screw dislocations incorporated into a mother phase in the early stages of crystallization provide a field for the emergence of the product B phase and a channel for the E solid-state transition. In addition, the polytypic transformation from a single-layer to a double-layer structure takes place during the crystallization of the E form. A similar polytypic transformation also occurs in petroselinic acid.

1. Introduction

Lipids and related long-chain compounds are closely connected with our daily life. Fatty acids and their derivatives are utilized as low materials or ingredients for soap, detergent, paint, medicine, cosmetics, food, and so on. Furthermore, some long-chain compounds are important from biological point of view. Phospholipids are one of the main constituents of biomembranes and their acyl chains form the hydrophobic portion of biomembranes. Triglycerides are also contained in plant and animal cells as storage fats.

One of the important characteristics of these compounds is the ability to form diversified aggregation states [1-5]. The hydrocarbon moieties of these compounds are able to take various conformations and lateral packings. Depending on surrounding condition and thermal history, a variety of crystal structures appear. The aggregation states of acyl chains have significant effects on functional activities of biomembranes[6]. Furthermore, the control of crystalline states is important for the quality control of many industrial products.

Many studies have been done about the structure and occurrence condition of each crystalline state of long-chain compounds. However, the detailed mechanism for the structure formation process of many solid states is still unclear. Usually

Y. Morishima · T. Norisuye · K. Tashiro (Eds.)
Molecular Interactions and Time-Space Organization in Macromolecular Systems
Proceedings of the OUMS '98, Osaka, Japan, 3-6 June, 1998

crystallization processes depicted with two steps, nucleation and successive growth. However, recent our study had suggested that this picture was too simple for some cases. Even if single crystal keeps the identical crystal morphology, there is a possibility that several types of structure changes take place in the internal parts, including solid-state phase transitions. We tried to confirm this phenomenon, but it was not long before we met an experimental difficulty; the structural changes were sometimes so small that their identifications were quite hard with ordinary experimental methods for the polymorphic study of lipids, such as X-ray diffraction and optical observation. To resolve this difficulty, we noticed the potential of Fourier-transform infrared (FT-IR) spectroscopy, a high-throughput optical system and highly efficient detectors, which makes IR measurement of a small single-crystal specimen easy and soon [7]. What is more, IR spectroscopy has the advantage of sensitivity to local structural changes of functional groups.

In this paper we mainly deal with the crystallization process of stearic acid, where two kinds of structural changes, polymorphic and polytypic transformations, occur. We show that crystal surface and defects have a significant influence on this phenomenon.

1. E→B Polymorphic Transition of Stearic Acid

Stearic acid has four polymorphs, A, B, C and E forms[1-3, 8-11]. Only the A form has the T// type subcell and the other three polymorphs form the O⊥ subcell. The C form is stable above 32°C, and the B form below 32°C [12-13]. The E form is metastable in the whole temperature region. Although the E form was first found by Holland and Nielsen in 1962 through a polarized IR spectroscopic study on single crystals [10-11], the existence of the E form was disregarded in most polymorphic studies of n-fatty acids owing to the lack of information about its occurrence condition and crystal structure. At the end of 1980s, we performed X-ray crystal structure analysis with accidentally obtained single crystals of the E form [14]. The obtained structure suggested the close relationship between the B and E forms. Except the conformation around carboxyl groups (contrary to the all-trans conformation of the E form, the B form has a gauche bond [8]), their crystal structures are almost the same with each other; the acyl chain inclines towards the b_S axis of the O⊥ subcell by 27° (the setting of the a_S, b_S and c_S axes is made in accordance with the orthorhombic polyethylene) as shown in Figure 1. Furthermore, the two polymorphs occur only by solution crystallization.

Because of their structural similarity, the E and B forms exhibit greatly intense X-ray reflections at almost the same 2θ angles and have the identical crystal morphology: a lozenge plate with an acute interedge angle of 74°. Therefore, the identification of the E form with optical observation and powder X-ray diffraction is almost impossible, which suggests that it might happen to regard E and B as one polymorph in the previous polymorphic studies. However, due to the difference in molecular conformation between B and E, there is a marked distinction in IR spectra. The OCO in-plane deformation δ(OCO) band appears at 648 cm^{-1} in B and 688 cm^{-1} in E. The progression bands of CH_2 wagging modes also exhibit marked differences.

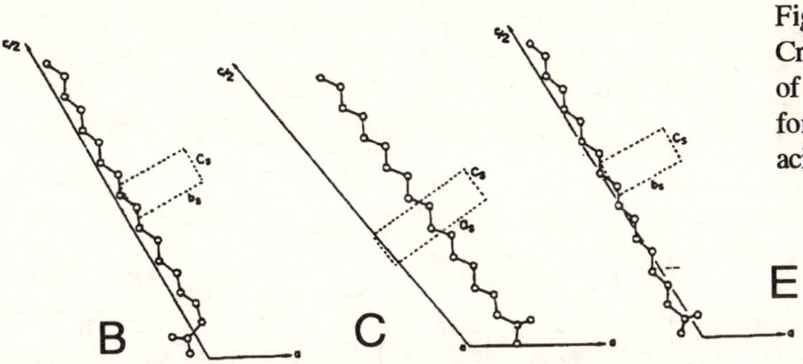

Figure 1. Crystal structures of B, C, and E forms of stearic acid

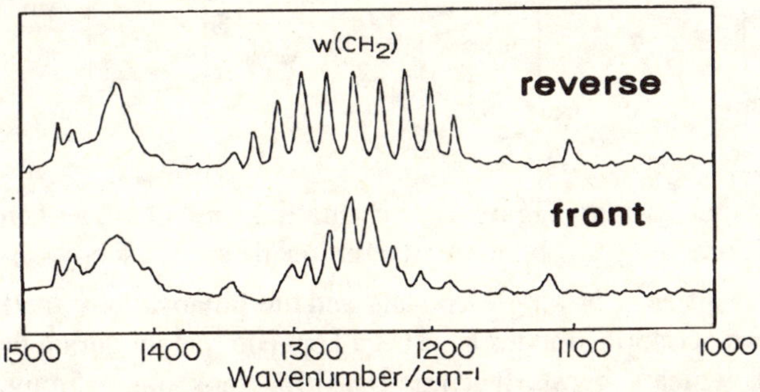

Figure 2. ATR-IR spectra taken at {0 0 1} faces of the BE type crystal.

Using these characteristic, we tried to reinvestigate the frequencies of occurrence of each polymorph. Single crystals were grown from a hexane solution by cooling and keeping it at the same temperature. Just after their appearance (within 1 hour), all single crystals exhibited the IR spectral feature of E, and then some single crystals began to show the spectral features both of B and of E. Here, we refer this type of crystal as BE type. After the occurrence of the BE type crystal, single crystals of the B form appeared in the solution. The ratio of the B crystal increased with time, and after 1 day all of the single crystals obtained were of the B form. We investigated the coexistence condition of the B and E polymorphs in the EB type crystals with ATR-IR method, which provides structural information about the surface region of several μ depth. The front face exhibited the spectral feature of B, whereas the reverse face showed those of E as shown in Figure 2. Namely, the B and E regions formed a layered structure in this crystal. The BE type crystal gradually transformed to the B form through a solid-state phase transition (Figure 3). However, pure single crystals of the E form did not transform spontaneously. It means that heterogeneous nucleation of the B form on the {0 0 1} faces of the E crystal is the necessary condition for the E→B solid-state transition. Once the B domain appeared on a single crystal in solution, the transition proceeded irrespective of the ambience; a single crystal of the BE type transformed to the B form in the atmosphere as well as in solution.

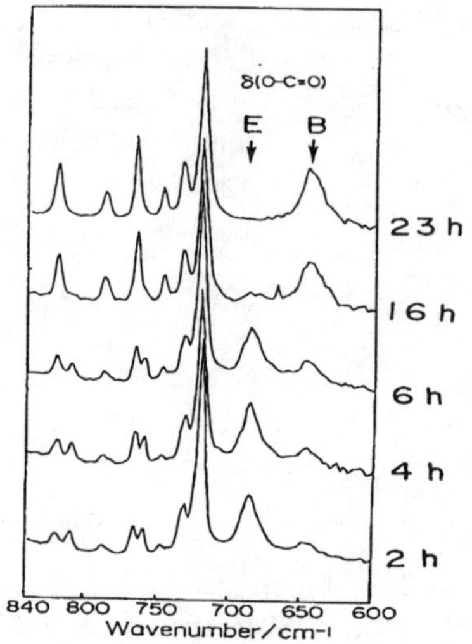

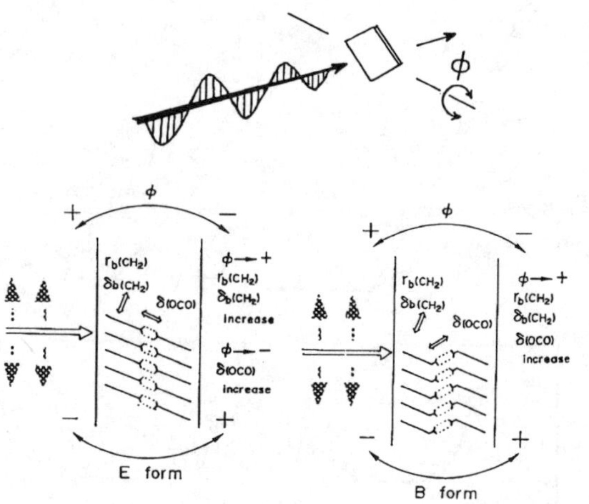

Figure 3. IR spectral changes during the E→B solid-state transition.

Figure 4. Schematic representation of the oblique transmission IR spectroscopy.

The crystal habit and transparency of single crystals and the polarizations in IR bands due to CH_2 rocking and scissoring modes have been kept completely during the E→B solid-state transition, which suggests that this transition took place through small localized structural changes. To confirm this expectation, the E→B solid-state transition was studied with IR oblique transmision method (Figure 4). IR spectra were measured at ϕ=0 and ±30° (ϕ=0°: normal incidence) and the plane of polarization was set parallel to the obtuse bisectrix. Figure 5 shows spectral changes of a BE-type crystal. In the starting crystal, the intensity of the δ(OCO) band was largest at ϕ=-30°, whereas that of the resultant B crystal was largest at ϕ=+30°. As E→B solid-state transition proceeded, the intensity of the band of B increased more pronouncedly at ϕ=+30° than at ϕ=-30°. On the other hand, the CH_2 rocking band at 720 cm^{-1} appeared most intense at ϕ=+30° for both the starting BE and the transformed crystal. From these spectral changes, it was clarified that only the portion near carboxyl group changed its arrangement during this transition.

A speculation can be made that the E→B transition proceeds in a bimolecular layer through the succession of this structure change from an all-*trans* to a bent conformation at the phase boundary between B and E, as depicted in Figure 6. However, the interlamellar interactions are so weak, compared with the intralamellar interactions, that the structure change on the crystal surface may not induce the phase transition in the interior. An *in-situ* observation of the growth processes clarified that not all single crystals transformed to the B form. In particular, most crystals generated at relatively moderate conditions remained as the E form.

Chemical etching showed a striking contrast between the two types of single crystals[16]. The single crystals, which had speedily transformed to B, exhibited many lozenge etch pits on the (0 0 1) plane, some of which reached the substrata of the crystals, while very few pits appeared in the crystals that did not convert to B.

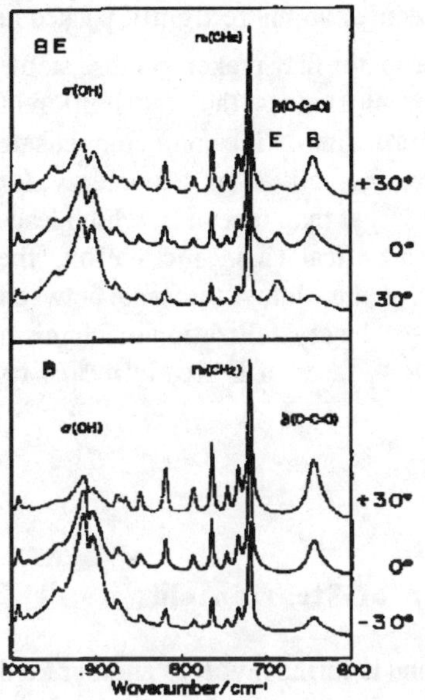

Figure 5. Oblique IR transmission spectral changes during the E→B solid-state transition.

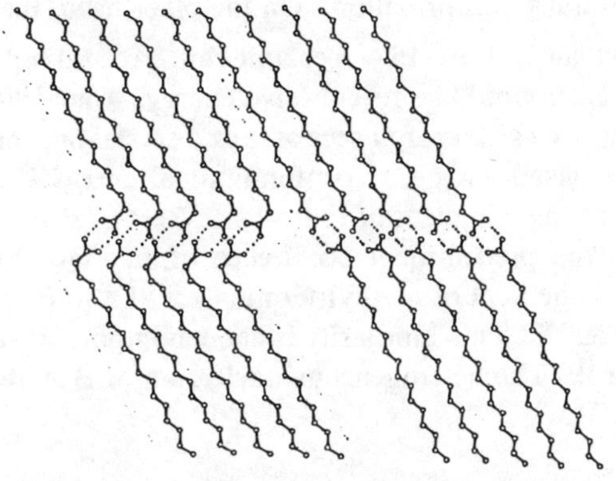

Figure 6. Schematic representation of the boundary between B and E.

The observation of growth spirals on the {0 0 1} faces of fatty acids' crystal suggested that the growth of the {0 0 1} faces was principally controlled by the spiral growth mechanism[17-21]. Taking this fact into consideration, one may explain the relationship of the striking contrast in chemical etching to the E→B transition, as follows. (1) Growth spiral steps due to screw dislocations are extremely long and durable, compared with the steps occurring in a 2-D nucleation growth mechanism; the long steps may catalyze the heterogeneous nucleation of the B form. (2) Since bimolecular layers form spiraling planes that are analogous to the floor of a modern multistoried parking garage with a slight upward gradient [22], the E→B transition is able to travel from a {0 0 1} face to the interior without interruption by driving on the ramp in a circle about a dislocation line.

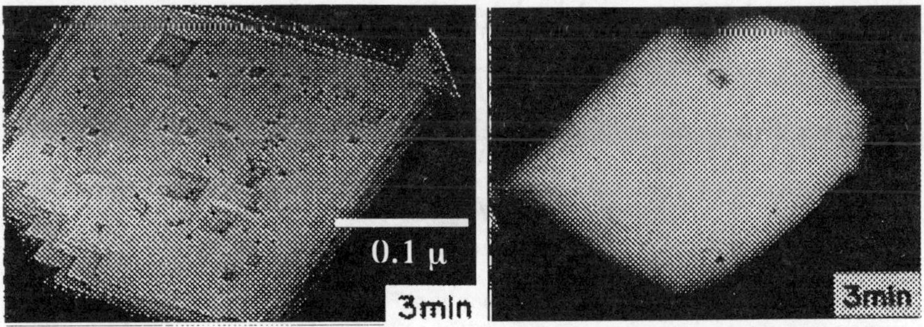

Figure 7. Optical micrographs showing the chemical etching process: (left) B form and (right) E form.

With respect to conformational energy, a molecule of the B form adopting a gauche conformation is less stable (by about 0.3 kcal/mol) than that of the E form taking an all-trans conformation. On the other hand, the molecules are more tightly packed in B than in E by 15Å^3 per unit cell. Accordingly, the factor that makes B more stable than E would be its cohesive energy. Under these circumstances, the contribution of the cohesive energy decreases as a crystal becomes small; finally it cannot compensate the disadvantage of conformational energy. Hence, the activation free energy for forming a nucleus of critical size ΔG^* of B would be larger than that of E, which leads to the predominant occurrence of E. Due to the identical chain inclination, the arrangement of methyl terminals at the {0 0 1} faces is the almost the same between B and E. This similarity is another important factor for the crystallization mechanism of B. The heterogeneous nucleation of B at the {0 0 1} faces of E greatly reduce its ΔG^*.

b. Orth II→Mon Polytypic Transformation of Stearic Acid

It has been found that the crystallization of the B and E forms involves another kind of structure change, a polytypic transformation [16]. Polytypism means a higher order structural difference caused by a different stacking sequence of layers [23]. In the long-chain compounds, polytypism has been observed in n-alkanes, n-alcohols, and fatty acids [24-28]. The B and E forms exhibit two polytypic structures: single-layer structure, Mon, and double-layer structure, Orth II [29-30] depicted in Figure 8. Contrary to the fixed chain inclination direction in Mon, the inclination direction is inverted at every stacking in Orth II. It has been found that the IR bands due to methyl C-H deformation $\delta(CH_3)$ modes are sensitive to the alternative chain stacking sequence [28]. Reflecting the difference in the CH_3-CH_3 stacking behavior at lamellar interfaces, Orth II showed its characteristic IR bands at 1382 and 1447 cm^{-1}, which were assigned to symmetric and antisymmetric methyl C-H deformation modes, $\delta s(CH_3)$ and $\delta a(CH_3)$.

Figure 8.
Two polytypic structures for fatty acids

Using the two marker IR bands, we monitored the polytypic conversion during the solution growth of the E form of stearic acid. Figure 9 shows an example, in which a saturated solution at 20°C was cooled to 16°C, where crystallization was initiated by stirring and kept quiescent for hours. The $\delta s(CH_3)$ and $\delta a(CH_3)$ bands were not found at the inception of crystallization, but gradually increased in intensity. Contrary to this spectral change, the spectral features of the E form in the other regions did not change. Namely, the first occurring single crystals were of Mon-type, and then gradually changed to the Orth II type. This transformation is due to the relative stability between polytypes. Orth II is thermodynamically more stable than Mon [12-13]. At the high supersaturation, the metastable Mon occurs for kinetic reasons. As the growth rate of Mon falls with decreasing solute concentration, the growth rate of Orth II surpasses that of Mon near the saturated concentration with respect to Mon. Accordingly, the single crystals begin to grow with the Orth II-type stacking.

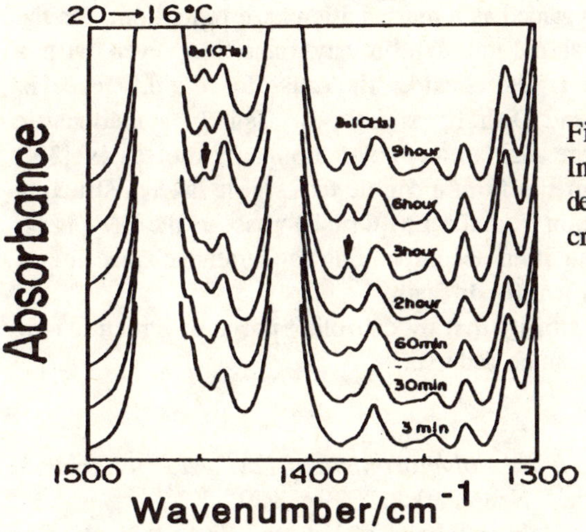

Figure 9.
Infrared spectral changes in methyl deformation regions of growing crystals.

C. Polytypic transformation and martensitic transition of petroselinic acid

As to cis-unsaturated fatty acids, the O⊥ subcell has been observed only in the low-melting (LM) phase of petroselinic acid (cis-13-octadecenoic acid) [31]. The methyl-sided chain of the LM phase of petroselinic acid is almost the same as the acyl chain of stearic acid E form in chain inclination and methyl terminal's packing. In solution crystallization, the LM phase also exhibits the polytypic transformation from Mon to Orth II.

Figure 10.
Resultant transformed regions of the martensitic transiton from the LM phase to the HM phase in petroselinic acid

0.5mm

Recently it was clarified that the polytypic structures of the LM phase controled not only thermodynamic propereties but also phase transitioin behavior[32]. A solid-state phase transition from the LM phase to the high-melting (HM) phase could be induced by a mechanical stress onto a single crystal. When a local stress was added onto a crystal crystal of Mon type (e.g., a press with the end of a needle), a part of single crystral transformed to the HM phase in a moment. The growth of the product phase was so fast that we could not follow it with the naked eye. The transformed region often displayed complex zigzag pattern as shown in Figure 10, but each transformed domainwas highly oriented: clear exitinction was observed under crossed Nicols. This transition of the LM phase can be regarded as a martensitic phase transition. On the other hand, the Orth II type crystal did not exhibit any transition, even with a mechanical stress enough to crack it. We consider the cause for the difference in transition behavior between Mon and Orth II, as follows. Usually a martensitic transition does not accompany large displacements of atoms or molecules [33]. However, large molecular rearrangements from a double to a single layered structure are required for the solid-state transition from the LM(Orth II) phase to the HM phase; under this condition a martensitic transition accompanying cooperative displacements of numerous molecules seems significantly difficult.

This phenomenon suggests a possibility that by conroling polytype structure new physical properties of a polymorph can be deduced.

References

[1] M. Kobayashi in *Crystallization and Polymorphism of Fats and Fatty Acids;* N. Garti and K. Sato Eds. (Marcel Dekker, New York, 1988) p. 139.

[2] K. Sato in *Crystallization and Polymorphism of Fats and Fatty Acids;* N. Garti and K. Sato Eds. (Marcel Dekker, New York, 1988) p. 227.

[3] *The Physical Chemistry of Lipids*, D. M. Small Ed. (Plenum, New York, 1986)

[4] J. M Hagemann *Crystallization andPolymorphism ofFats andFatty Acids;* N. Garti and K. Sato Eds. (Marcel Dekker, New York, 1988) p. 9.

[5] M. Kobayashi and F. Kaneko, J. Dispersion Sci. Tec. 10 (1989) 319.

[6] B. L. Silver, *The Physical Chemistry of Membranes* (Solomon, New York,1985)

[7] P. R. Griffith and J. A. de Haseth, *Fourier Transform InfraredSpectroscopy* (Wiley, New York, 1986).

[8] M. Goto and E. Asada, Bull. Chem. Soc. Jpn., 51 (1978) 2456.

[9] V. Marta, G. Celotti, R. Zanetti, and A. F. Martelli, J. Chem. Soc. B (1971) 548.

[10] R. F. Holland and J. R. Nielsen, J. Mol. Spectrosc., 9 (1962) 436.

[11] R. F. Holland and J. R. Nielsen, Acta Crystallogr., 16 (1963) 902.

[12] K. Sato, M. Kobayashi, H. Morishita, J. Crystal Growth, 87 (1988) 236.

[13] K. Sato and M. Kobayashi, Crystal, 13 (1991) 65.

[14] F. Kaneko, M. Kobayashi, Y. Kitagawa and Y. Matsuura, Acta Cryst. C46 (1990) 1490.

[15] F. Kaneko, T. Simofuku, H. Miyamoto, M. Kobayashi, and M. Suzuki, J. Phys. Chem. 96 (1992) 10554.

[16] F. Kaneko, H. Sakashita, M. Kobayashi and M. Suzuki, J. Phys. Chem. 98 (1994) 3801.

[17] A. R. Verma and P. Reynolds, Proc. Phys. Soc(London) B66 (1953) 414.

[18] N. G. Anderson and I. M. Dawson, Proc. Phys. Soc(London) A218 (1953) 255.

[19] K. Sato and M. Okada, J. Crystal Growth, 42 (1977) 264.
[20] K. Sato, Jpn. J. Appl. Phys., 18 (1979) 1019.
[21] R. Boistelle, in *Crystallization and Polymorphism of Fats and Fatty Acids;* N. Garti and K. Sato Eds. (Marcel Dekker, New York, 1988) p. 189.
[22] J. Weertman and J. R. Weertman *Elementary Dislocation Theory* (Oxford, New York, 1992)
[23] K. Sato and N. Garti in *Crystallization and Polymorphism of Fats and Fatty Acids;* N. Garti and K. Sato Eds. (Marcel Dekker, New York, 1988) p. 3.
[24] R. Boistelle, B. Simon and G. Pepe, Acta Cryst., B32 (1976) 1240.
[25] A. R. Verma, Proc. Roy. Soc. (London) A28 (1955) 34.
[26] S. Amelinckx, Acta Cryst. 9 (1956) 16.
[27] D. Aquilano, J. Crystal Growth 37 (1977) 215.
[28] M. Kobayashi, T. Kobayashi, Y. Itoh, Y. Chatani and H. Tadokoro, J. Chem. Phys. 72 (1980) 2024.
[29] F. Kaneko, H. Sakashita, M. Kobayashi, Y. Kitagawa and Y. Matsuura, Acta Crystallogr., C50 (1994) 245.
[30] F. Kaneko, H. Sakashita, M. Kobayashi, Y. Kitagawa and Y. Matsuura, Acta Crystallogr., C50 (1994) 247.
[31] F. Kaneko, M. Kobayashi, Y. Kitagawa, Y. Matsuura, K. Sato and M. Suzuki, Acta Crystallogr. C48 (1992) 1054.
[32] F. Kaneko, M. Kobayashi, K. Sato and M. Suzuki, J. Chem. Phys, 101 (1997) 285.

Fluorescence Studies on Molecular Interaction and Dynamics of Liquid Crystalline Polymers

Kazuyuki Horie*, Hsu-Wen Huang, Masaki Shimada

Department of Chemistry and Biotechnology, Graduate School of Engineering, University of Tokyo, 7-3-1 Hongo, Bunkyo-ku, Tokyo 113-8656, Japan

Abstract. Fluorescence measurements on several main-chain biphenyl-type liquid crystalline polyesters were carried out to investigate the temperature dependence of molecular interaction between mesogenic groups and the dynamics around the mesogenic groups in the time range of nano-seconds. A blue shift of fluorescence during heating was observed for BB-6 with 4,4'-biphenyldicarboxylate mesogen, while a red shift with strong excitation wavelength dependence was observed for PB-10 with 4,4'-dicarbonyloxybiphenyl mesogen. The difference is attributed to the difference in the temperature dependence of packing patterns for these polyesters. Fluorescence depolarization in a liquid crystalline polymer was measured for the first time by using perylene in BB-5 with 4,4'-biphenyldicarboxylate mesogen. Rotational diffusion coefficient increases gradually with the increase in temperature and jumps at the liquid crystal-isotropic transition temperature.

Introduction

It is well known that fluorescence is a very useful tool for investigating the interaction and dynamics of polymer chains in solution as well as in the solid state. When we apply the fluorescence technics to liquid crystals (LC) [1] and LC polymers [2,3], important information which we can obtain is (1) the appearance of new fluorescent components, (2) the shift of fluorescence peaks, and (3) the fluorescence anisotropy ratio.

As to the first category, the observation of charge-transfer (CT) fluorescence from a LC polymer of rigid-rod aromatic polyester composed of pyromellitic and biphenyl groups with long alkyl side chains (B-C16) [4,5] has elucidated the spatial packing structure of aromatic main chains within a layer where biphenyl moieties lie adjacent to the pyromellitic ester groups. However, in a lyotropic polyester of poly(sulfo-1,4-phenylene 3'-nitroterephthalate) (PSPNT), fluorescence study [6] has

Y. Morishima · T. Norisuye · K. Tashiro (Eds.)
Molecular Interactions and Time-Space Organization in Macromolecular Systems
Proceedings of the OUMS '98, Osaka, Japan, 3-6 June, 1998

suggested that the microstructure of the lyotropic LC system involves a parallel alignment of polymer chains with interaction of structurally-similar chromophores being favored over interactions between different aromatic chromophores.

The fluorescence measurement of a thermotropic LC polyester from biphenyl-4'4-dicarboxylic acids and alkanediol (BB-n) [7] shows blue shift of its peak during heating without giving excitation wavelength dependence. The blue shift of fluorescence peak during heating is also observed for a low-molecular-weight LC, 4-octyl-4'-cyanobiphenyl [1]. However, fluorescence measurements of another thermotropic LC polyester prepared from 4,4'-dihydroxybiphenyl and aliphatic dibasic acids containing n methylene units (PB-n) [8,9] showed marked excitation wavelength dependence of fluorescence peaks of its ground-state complexes and their red shift during heating. The marked excitation wavelength dependence of the fluorescence spectra would be attributed to disordered crystalline nature of PB-n.

Fluorescence anisotropy ratio in a LC system has been measured for the first time by using perylene as a fluorescent probe in BB-5 [10]. The rotational diffusion coefficient, D_r, increases gradually with increasing temperature in the crystalline phase, becomes to be 5 - 10^7 s^{-1} in the S_{CA} phase at 172 - 200 °C, then jumps to 1 - 10^8 s^{-1} in the isotropic phase.

In the present paper, the second and third topics, i.e., the shift of fluorescence peaks and fluorescence anisotropy ratio of LC polymers are discussed in detail.

2. Blue Shift of Fluorescence during Heating of a Liquid-Crystalline Polyester (BB-6)

A main-chain liquid-crystalline polyester with 4,4'-biphenyldicarboxylate mesogens (BB-6) is selected as the first example to study the change in molecular interaction during heating. The crystalline and LC phases of BB-6 were studied extensively by many researchers [11-13]. According to these studies, BB-6

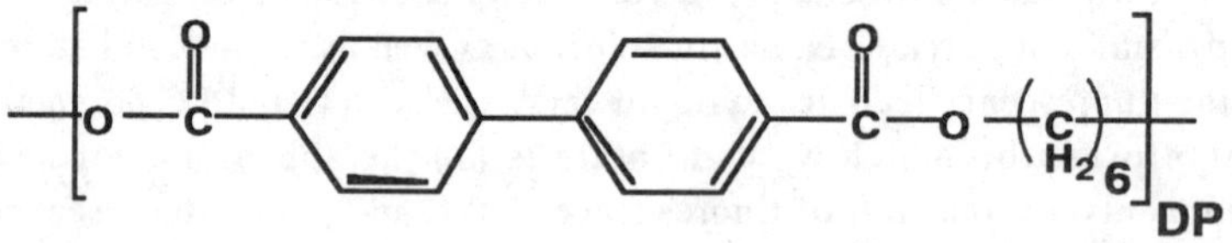

BB-6

was confirmed to show a ordered K_β crystalline phase, a random K_γ crystalline phase, and a stable S_A phase during heating. However, the situation of degree of overlap between mesogens in LC layer and the progress of phase transition in crystalline mixture are still unknown. Therefore, fluorescence method is introduced to elucidate the above microstructure change.

Figure 1a shows the fluorescence and its excitation spectra of BB-6 at room temperature. BB-6 shows fluorescence at 384 nm (a_1,a_2), which is almost identical with the fluorescence of a crystal of its model compound, 4,4'-biphenyldicarboxylic acid dimethyl ester (BDDE) (378 nm). This indicates that the intermolecular ground-state complex is formed between 4,4'-biphenyldicarboxylate mesogens and shows fluorescence at 384 nm. But the arrangement between mesogens of the fluorescence at 384 nm for BB-6 is still unknown, i.e., the fluorescence at 384 nm is originated from the overlapping arrangement between biphenyl groups (Figure 1b) or the

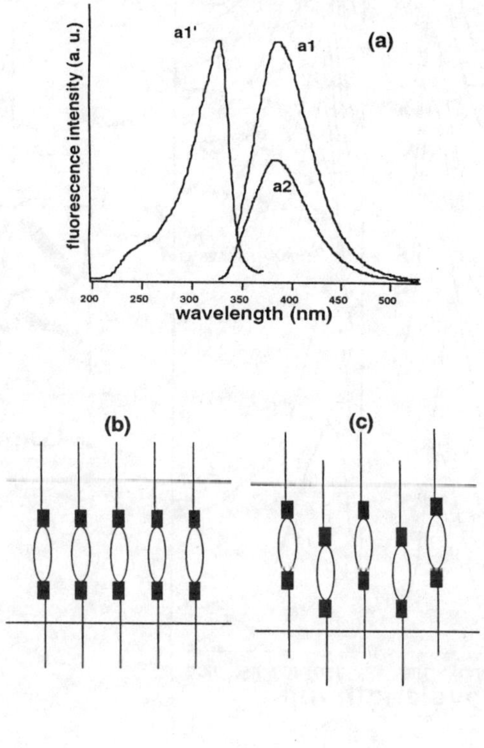

Figure 1. (a) The fluorescence and its excitation spectra of BB-6; a_1 and a_2 are the fluorescence spectra excited at 330 and 300 nm, respectively; a1' is the fluorescence excitation spectra monitored at 384 nm. (b) Overlapping arrangement between biphenyl groups. (c) Overlapping arrangement between biphenyl groups and carbonyl groups.

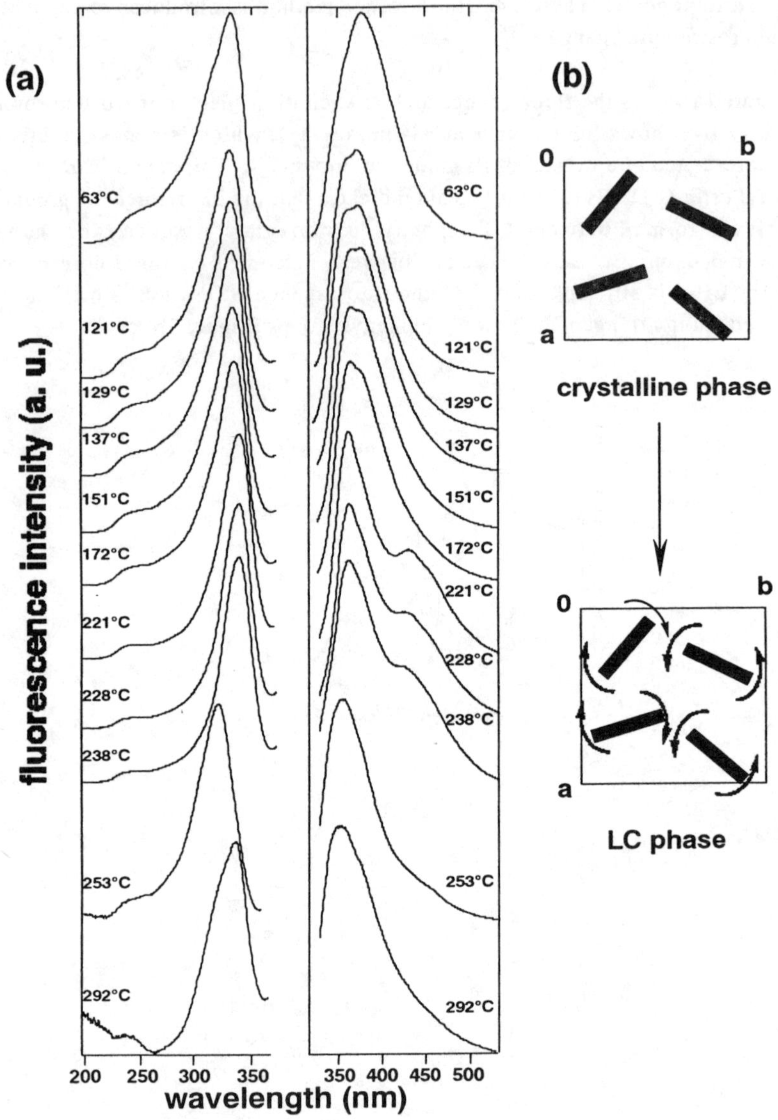

Figure 2. (a) Fluorescence (right) and its excitation spectra (left) of BB-6 during heating. Fluorescence spectra are all excited at 310 nm and excitation spectra are all monitored at 384 nm. (b) Projection view of the unit cells onto the ab plane in crystalline phase and LC phase.

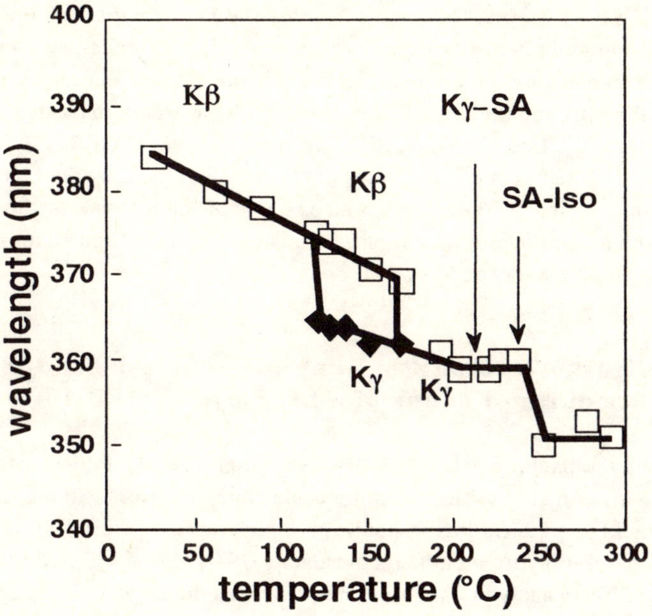

Figure 3. The change in fluorescence peak wavelength during heating for BB-6 (excited at 310 nm). ◆: the lower part of fluorescence peak wavelength of BB-6 during crystalline mixture phase transition process.

overlapping arrangement between biphenyl groups and carbonyl groups (Figure 1c). Based on to the crystal structure of BDDE [14], it can be said that the fluorescence of BB-6 is originated from the overlapping arrangement between biphenyl groups and carbonyl groups as depicted in Figure 1c.

Figure 2a shows the fluorescence and its excitation spectra of BB-6 during heating. Fluorescence peak wavelength of BB-6 shifts to shorter wavelength from several crystalline phases to isotropic phase. This would be because the packing pattern is almost identical during heating, and the increase in thermal fluctuation makes the rotational fluctuation increase, then causes the decrease in molecular interaction between mesogens (Figure 2b). Thus, the electronic state of intermolecular ground-state complex becomes more similar to that of monomer. This is the reason why the fluorescence peak wavelength of BB-6 shows a blue shift phenomenon during heating.

The temperature dependence of fluorescence peak wavelength for BB-6 during heating is depicted in Figure 3. The fluorescence peak wavelength shows gradual shift with increasing temperature. This indicates that the LC transition occurs first at the microscopic change in mobility between mesogens then extends to the entire phase around phase transition temperature. Moreover, the fluorescence peak wavelength shows separation around the crystalline mixture phase. This indicates that the K_β phase exists initially, then the K_γ phase begins to appear at around 121 °C, and the K_β phase decreases but the K_γ phase increases gradually. Finally, only the K_γ phase is left before LC transition temperature. This is an example of fluorescence method providing the information about the progress of phase transition in crystalline mixture which has not been provided by X-ray analysis.

3. Excitation Wavelength Dependence and Red Shift of Fluorescence during Heating of a LC Polyester (PB-10)

Since the packing pattern of BB-6 is almost identical during heating, it is interesting to investigate the change in molecular interaction of LCP which shows the change in packing pattern during heating. Thus, a main-chain liquid-crystalline polyester with 4,4'-dihydroxybiphenyl mesogens (PB-10) was selected for this study. From DSC and X-ray analysis, the abrupt decrease was observed in layer spacing at K-S_H transition temperature.

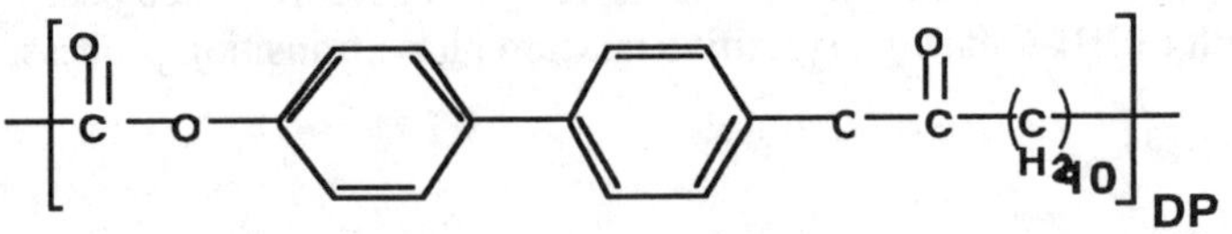

PB-10

It was proposed that the molecular arrangement is fully overlapping arrangement between biphenyl groups in crystalline phase, but changes to the overlapping arrangement between biphenyl groups and carbonyl groups in S_H phase, by assuming all-trans conformation of the alkyl spacer [15]. Thus, it is necessary to investigate the change in degree of overlap itself between mesogens during heating by fluorescence method.

Figure 4 shows the fluorescence (right part) and its excitation spectra (left part) of PB-10 at crystalline phase [8]. Fluorescence spectra excited at 260-300 nm (A, a) show fluorescence only at 343 nm, which is demonstrated to be due to the monomer fluorescence. But the fluorescence excited at 320-400 nm (b-i) show wavelength shift to red with increasing excitation wavelength. The shape of fluorescence excitation spectrum of monomer (a') is also different from those monitored at longer wavelength (b', c'). This indicates that the fluorescence spectra excited above 320 nm are due to the intermolecular ground-state complexes, but are not due to the excimer. Moreover,

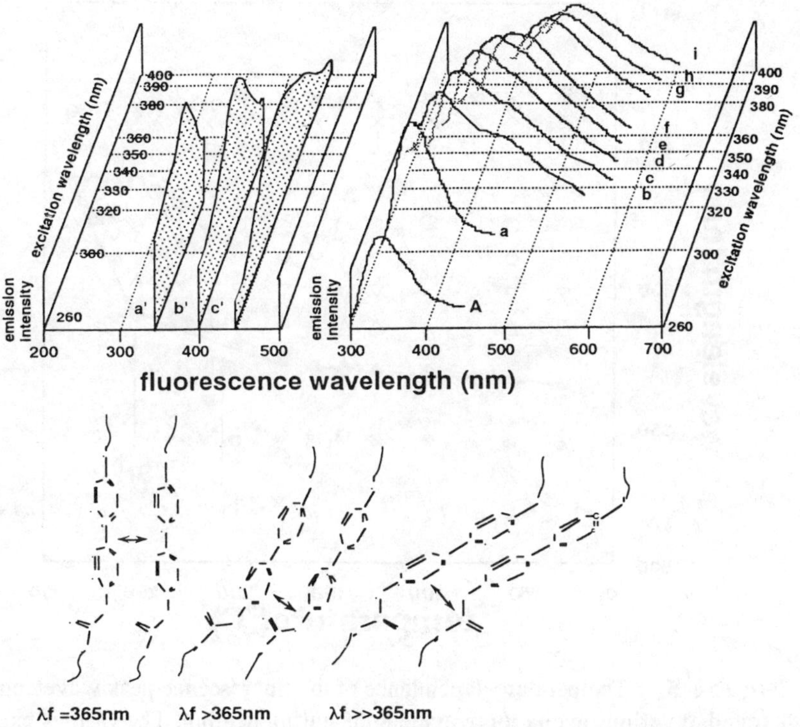

Figure 4. The fluorescence (right part) and its excitation spectra (left part) for PB-10 cooled from 270°C and measured at 26°C. A, a, b, c, d, e, f, g, h, and i are the fluorescence spectar excited at 260 nm, 300 nm, 320 nm, 330 nm, 340 nm, 350 nm, 360 nm, 380 nm, 390 nm, and 400 nm, respectively. a', b', and c' are the fluorescence excitation spectra monitored at 343 nm, 400 nm, and 450 nm, respectively.

the fluorescence peak wavelength excited at 320 nm of PB-10 (365 nm) is almost in accord with the fluorescence of intermolecular ground-state complex of 4,4'-diacetoxybiphenyl (DABP) crystal (362 nm) which indicates a fully overlapping arrangement between biphenyl groups [16]. The fluorescence peak wavelength longer than 365 nm is demonstrated to be due to the interaction between the biphenyl groups and carbonyl groups as illustrated in the lower part of Figure 4. This kind of excitation wavelength dependence of fluorescence also implies the existence of various aggregation structures formed by the different degree of overlap between mesogens. To the best of our knowledge, generally, amorphous polymer itself usually shows one

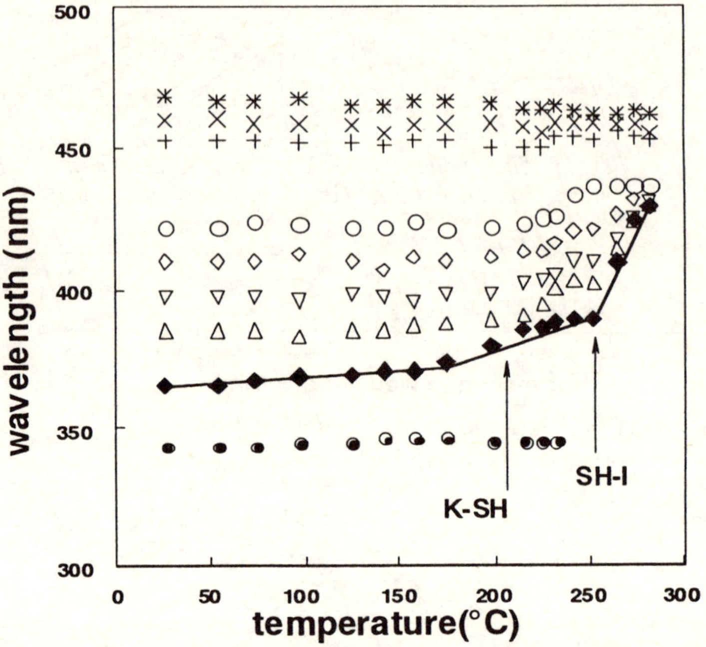

Figure 5. Temperature dependence of the fluorescence peak wavelengths excited at various excitation wavelengths during heating. The various excitation wavelengths are as follows. (●): 300 nm; (◆): 320 nm; (△): 330nm; (▽): 340 nm; (⊕): 350 nm; (○) 360 nm; (+): 380 nm; (×): 390 nm; (*): 400 nm.

fluorescence peak wavelength, when the excitation wavelength is changed little by little, except for the red edge effect of a twisted intramolecular charge transfer (TICT) probe [17] in poly(hydroxyethyl methacrylate). The fluorescence of red edge effect is originated from the different conformations of TICT probe, but is not originated from the polymer itself. Thus, the strong excitation wavelength dependence of PB-10 would be a novel discovery for polymer fluorescence, which may be due to the formation of a disordered crystal.

Figure 5 shows the temperature dependence of fluorescence peak wavelengths excited at various wavelengths during heating. The fluorescence peak wavelength of monomer (•) and those excited at longer wavelength (+, ×, ∗) do not show wavelength shift during heating. But the fluorescence peak wavelengths related to the degrees of overlap between mesogens (♦, Δ, ∇, ⊕, and ○) show red shift during heating and converge to a broad structureless emission peak centered at 429 nm in the isotropic phase. Moreover, fluorescence peak wavelength corresponding to the fully-overlapping arrangement between biphenyl groups (♦) shows break before K-S_H transition. This implies that the temperature effect initially makes the phase transition start in a local microstructure change (molecular interaction between mesogens) then extended to the macrostructure change (the entire phase transition). The red shift of fluorescence peak wavelengths of PB-10 to longer wavelength during heating is totally different from the blue shift of BB-6. The red shift of fluorescence is caused by the decrease in the degree of overlap between mesogens, which induces the CT interaction between biphenyl groups and carbonyl groups.

4. Fluorescence Depolarization Study on a Liquid-Crystalline Polyester (BB-5)

Fluorescence depolarization measurements of perylene in a thermotropic LC polyester (BB-5) was carried out for evaluating dynamics around mesogen molecules in the LC polymer. BB-5 shows crystalline (K) to liquid crystal (S_{CA}) transition at 175 °C and from S_{CA} to isotrapic (I) transition at 211 °C during heating [12]. In cooling process it shows hysterisis.

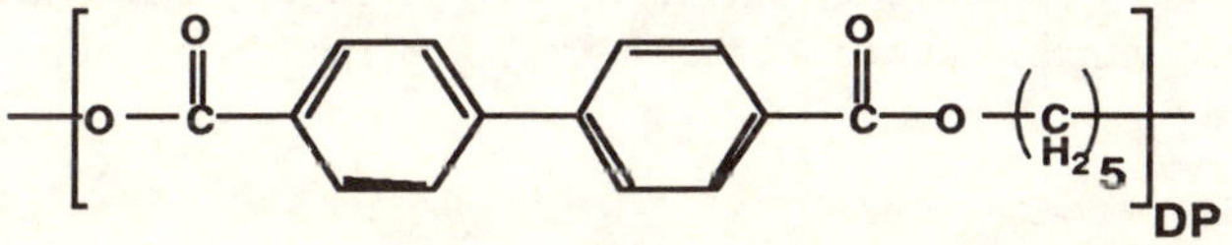

BB-5

Perylene (6.5 × 10^{-3} mol/kg) was molecularly dispersed in a BB-5 film by spin-coating from chloroform solution. The film samples were vacuum-dried at room temperature for a day and annealed at the isotropic temperature before fluorescence measurements. The transition temperatures were not influenced by the presence of perylene molecules. Perylene in a BB-5 film shows monomer fluorescence peaks at 456 and 482 nm and a shoulder at 516 nm when excited at 420 nm. Its fluorescence intensity decreases gradually with increasing temperature up to 220 °C but peak wavelengths do not shift. As the profiles of these spectra are very similar to those of perylene in a dilute solution, we can assume the absence of any special complex formation between perylene and mesogenic biphenyldicarboxylate moiety.

In fluorescence depolarization measurements of liquid crystalline samples, the correction for scattering light from the samples is needed. Fluorescence anisotropy ratio, r, by steady-state measurements is given by

$$r = (I_{VV} - GI_{VH}) / (I_{VV} + 2GI_{VH}) \quad (1)$$

where, I is the fluorescence intensity with first and second subscripts referring to the vertical or horizontal orientation of the excitation and observation polarizer, respectively, and G is the correction factor for the anisotropy of apparatus. When the sample scatters light to some extent, a correction for r can be done by measuring the scattering anisotropy ratio, r_{ex} and r_{em}, for the condition of irradiating and monitoring at the same wavelength each corresponding to excitation and emission of the sample, respectively [18]. The corrected fluorescence anisotropy ratio, r_c, was then obtained from

$$r_c = r / (r_{ex}\, r_{em})^{1/2} \quad (2)$$

Figure 6 shows the change in r_c of perylene fluorescence in BB-5 excited at 420 nm and emitting at 480 nm during heating. The value of r_c decreases gradually during heating in the crystalline phase suggesting the gradual increase in local molecular motion, then r_c becomes constant in S_{CA} state, and finally shows an abrupt drop in the isotropic mobile phase. Similar change due to the occurrence of local rotational fluctuation of mesogen moieties of BB-5 during heating has also been observed as a blue shift of fluorescence peak wavelength of BB-5 itself during heating [7].

The rotational diffusion coefficient, D_r, for pyrelene in BB-5 film was calculated from eq. (3) with the values of r_c, the fluorescence lifetime, τ, and $r_o = 0.34$ [19], and is shown in Figure 7.

$$r_o / r_c = 1 + 6D_r\tau \quad (3)$$

The values of D_r increase gradually with increasing temperature in the crystalline phase, suggesting that microscopic rotational motion around mesogen moieties seems to be gradually liberated with increasing temperature even in the crystalline phase below K-S_{CA} transition (172 °C), while macroscopic phase transition occurs cooperatively at the transition temperature. D_r becomes to be 5×10^7 s^{-1} in the S_{CA} phase at 172 - 200 °C, and jumps to 1×10^8 s^{-1} in the isotropic phase. In such a way, we can obtain from fluorescence studies important information on local dynamics around mesogen groups of LC polymers.

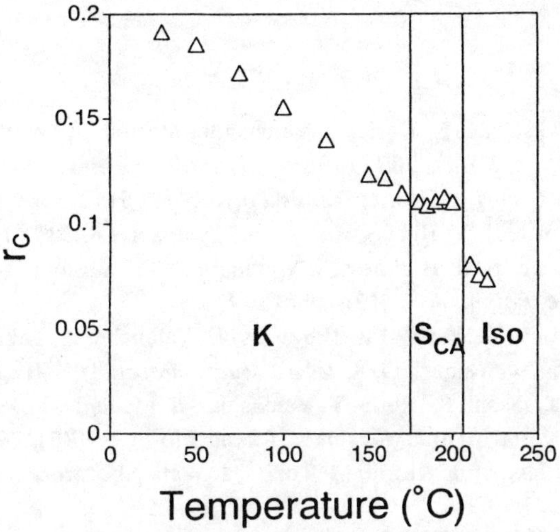

Figure 6. Change in corrected fluorescence anisotropy ratio, r_C, for perylene in BB-5 during heating.

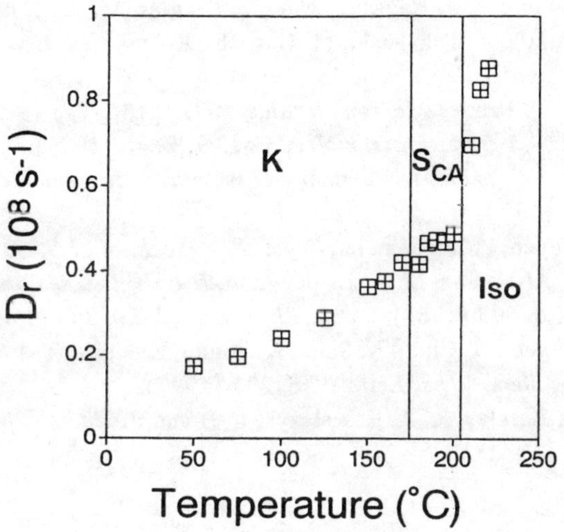

Figure 7. Change in rotational diffusion coefficient, D_r, of perylene in BB-5 during heating.

References

1. H. Takahashi, K. Horie, T. Yamashita, M. Sone, J. Watanabe, *Liquid Crystals,* **2 1**(4), 505 (1996).
2. H.W. Huang, K. Horie, *Trends Polym. Sci.,* **5** (12), 407 (1997).
3. K. Horie, H.W. Huang, *Macromol. Symp.,* **1 1 8**, 229 (1997).
4. M. Sone, B. R. Harkness, J. Watanabe, T. Yamashita, T. Torii, K. Horie, *Polym. J.,* **2 5**(9), 997 (1993).
5. M. Sone, T. Wada, B.R. Harkness, J. Watanabe, H. Takahashi, H.W. Huang, T. Yamashita, K. Horie, *Macromolecules,* **3 1**(25) (1998).
6. H. Takahashi, K. Horie, T. Yamashita, S. Machida, D.T.B. Hannah, D.C. Sherrington, *Macromol. Chem. Phys.,* **1 9 7**(9), 2703 (1996).
7. H.W. Huang, K. Horie, M. Tokita, J. Watanabe, *Macromol. Chem. Phys.,* in press.
8. H.W. Huang, K. Horie, T. Yamashita, S. Machida, M. Sone, M. Tokita, J. Watanabe, Y. Maeda, *Macromolecules,* **2 9**(10), 3485 (1996).
9. H.W. Huang, K. Horie, M. Tokita, J. Watanabe, *Polymer,* in press.
10. M. Shimada, H. W. Huang, S. Machida, K. Horie, J. Watanabe, IUPAC MACRO98 Gold Coast, preprints, p. 777 (1998).
11. J. Watanabe, M. Hayashi, *Macromolecules,* **2 1**, 278, 4083 (1988).
12. M. Tokita, T. Takahashi, M. Hayashi, K. Inomata, J. Watanabe, *Macromolecules,* **2 9**, 1345 (1996).
13. X. Li, F. Brisse, *Macromolecules,* **2 7**, 7718, 7725 (1994).
14. X. Li, F. Brisse, *Macromolecules,* **2 7**, 2276 (1994).
15. W. R. Krigbaum, J. Watanabe, T. Ishikawa, *Macromolecules,* **1 6**, 1271 (1983).
16. W. Thyen, P. Zugenmaier, *Cryst. Res. Technol.,* **2 9**, 591 (1994).
17. K. A. Al-Hassan, M. A. El-Bayoumi, *J. Polym. Sci. Polym.Phys. Ed.,* **2 5**, 495 (1987); *Chem. Phys. Lett.,* **1 5 0**, 344 (1988).
18. S. Tsuneda, T. Endo, K. Saito, K. Sugita, K. Horie, T. Yamashita, T. Sugo, *Chem. Phys. Lett.,* **2 7 5**, 203 (1997).
19. M. D. Barkley, A. A. Kowalczyk, L. Brand, *J. Chem. Phys.,* **7 5**, 3581 (1981).

Molecular Rheology of Glassy Polymers

K. Osaki and T. Inoue

Institute for Chemical Research, Kyoto University, Uji, Kyoto 611-0011, Japan
osaki@scl.kyoto-u.ac.jp

Abstract. Simultaneous measurements of stress and birefringence in oscillatory deformation were revealed to provide information on the deformation of polymer chains and origin of stress in glassy state. The stress was composed of the glassy stress similar to that of ordinary glass-forming liquids and the rubbery stress due to the entropic force of chains. The relaxation behavior of these stress components were discussed in details.

1. Introduction

A big subject of molecular rheology is to clarify how the chain molecules deform when the material is deformed, how the stress is originated, and how the stress relaxes as the deformed molecules recover equilibrium. The characteristic relaxation spectrum of modulus, including the glassy, glass-to-rubber transition, rubbery plateau, and terminal flow zones, of flexible polymers implies various stages of molecular deformation as well as of origin of stress. In soft states of high temperatures or long times, rubbery and terminal flow zones, the molecular chain may be modeled as a Gaussian chain that exerts tensile force of entropic nature when two ends are fixed in space. The Gaussian chain is a model in which details of the chain shape has been smeared out in a very short time leaving only the gross contour. Due to the simple nature of Gaussian chain, the essential features of molecular rheology, including that in large deformations, have been formulated up to a realistic level well compared to observed results.[1)]

The modulus and its relaxation at shorter times are apparently similar to those of materials of low molecular weight in supercooled state. The relaxation of high glassy modulus is associated with dielectric segmental relaxation. Thus the modulus must be related to the details of deformation at smaller portions of chain, probably represented by the monomer orientation.

The idea of the present study is to measure the monomer orientation as anisotropy of refractive index at various stages of stress relaxation. The birefringence is proportional to the stress for materials modeled with Gaussian chains. Any other mode of orientation will be associated with the stress in another manner. By comparing the stress and birefringence, we have been able to see how the monomer orientation is determined on instantaneous macroscopic deformation and how the stress relaxes as the orientation of chain axis and the rotational orientation of monomers equilibrate. The chain axis orientation becomes detectable at the chain length approximately corresponding to the Kuhn statistical segment.

2. Stress and birefringence in the glassy and glass-to-rubber transition zones

We measure the tensile stress and the birefringence of a film in oscillatory elongation. The frequency dependence of complex Young's modulus, $E^* = E'(\omega) + iE''(\omega)$, is essentially the same for any polymer provided the comparison is made at appropriate reference temperature for each polymer. The strain-optical ratio, $O^* = O'(\omega) + iO''(\omega)$, defined as the complex ratio of birefringence to strain is proportional to E^* only at low frequencies or high temperatures. In the glassy and glass-to-rubber transition zones, O^* varies in three typical manners, classified by how each component changes signs as frequency changes. These features of stress and birefringence can be described with a modified stress-optical rule (MSOR).[2)]

$$E^* = E_R^* + E_G^*, \qquad O^* = C_R E_R^* + C_G E_G^*$$

Here we defined two complex functions E_R^* and E_G^*, respectively associated with relatively long and short relaxation times. C_R and C_G are stress-optical coefficients for respective stress components. C_R is the ordinary stress-optical coefficient in the soft states and its relation to the molecular structure is well known. Three types of frequency dependence of O^* is accounted for by magnitudes and signs of two coefficients.[3)]

The component function $E_G^*(\omega)$ was essentially the same for many polymers and for supercooled materials provided that an appropriate reference temperature was chosen for each material. $E_R^*(\omega)$ was fitted well with the prediction of the Rouse model and the Rouse segment determined from the instantaneous value of R-modulus through a relation $M_s = 3\rho RT/E_R'(\infty)$ was close to the Kuhn statistical segment.[4, 5)]

3. Molecular strain as origin of stress for instantaneous deformation in the glassy zone

The birefringence is determined by the anisotropy of polarizability, the orientation, and the number of monomers. The high frequency limit of the component functions, $O_R'(\infty)$ and $O_G'(\infty)$, would represent the instantaneous orientations due to R- and G- orientation modes, say. For flat monomers the orientation may be defined by assigning the chain axis orientation and the rotation angle around the axis. Detailed estimation for various polymers revealed that the R mode corresponded to the chain orientation and the G mode to the monomer rotation.[6, 7] The $O_R'(\infty)$ was well evaluated by assuming a quai-affine orientation of the chain axis, which would be the case if the chain contour was embedded in the macroscopic body. The rotation angle around the axis may depend on the monomer shape. Monomers bulky sideways were more likely to rotate by itself according to the sample deformation disregarding the chain connectivity, or distorting the axis orientation to some degree. Thinner monomers respected the axis orientation and rotated about the oriented axis.[7]

One can deduce from these results that $O_R'(\infty)$ is equal to 3/5 of the intrinsic birefringence, the birefringence corresponding to the full extension of chain.[8] On the other hand, the stress-optical coefficient, $C_R = O_R'(\infty)/E_R'(\infty)$, represents the anisotropy of the Rouse segment and will be smaller for a more flexible chain even if the monomer anisotropies are the same for two polymers. This can be tested through comparison of atactic and syndiotactic polystyrenes.[9]

4. Elongation at constant rate

The stress and birefringence in elongation at constant rate can also be transformed to the component stresses through the MSOR.[10] The R-component of stress varied according to the linear viscoelasticity law and then diverged when the Hencky strain approached 2. This is common to model theories for flexible polymers unless any particular mechanism for nonlinearity is deviced. The G-component stress exhibited an overshoot when the rate of elongation was high and the steady value was much lower than estimated by the linear viscoelasticity. The strong strain softening seems to be common to glassy materials.

5. Properties of low M polystyrene

For entangled polymers, the R-modulus varies from $E_R'(\infty)$ to the entanglement modulus E_N in the glass-to-rubber transition zone. When the molecular weight was not high enough to form the entanglement, the $E_R^*(\omega)$ could be described by the Rouse model. The only one adjustable parameter was the unit of the frequency scale or the friction coefficient of the segment. Other quantities were completely determined from the molecular weight and the segment molecular weight, M_S, determined for polymers with high molecular weights.

As the molecular weight still decreased down to about $2M_S$, the instantaneous modulus for the R component, $E_R'(\infty)$, became lower than that estimated from the M_S value. The R component of stress almost vanished at the molecular weight comparable to M_S. This result seems to indicate that the chain axis or the chain connectivity becomes significant at a molecular weight of the order of M_S as far as the monomer orientation in deformed material is concerned.[11)]

6. Concluding remark

We may say that we know the monomer orientation in deformation, the origin of stress, and the relaxation process as far as the chain axis is concerned. The remaining part of stress is associated with the monomer rotation about the oriented chain axis and it is similar to that of ordinary glassy materials. The origin of stress is definitely the perturbed atomic distances caused by the deformation. One may be able to guess the orientation of monomers in various stages of relaxation. But does this help theoretical derivation of the high modulus and the characteristic relaxation spectrum ?

References

1) See for example, D. Doi and S. F. Edwards, *TheTheory of Polymer Dynamics*, Clarendon, 1986.
2) T. Inoue, T. Okamoto, and K. Osaki, *Macromolecules*, **24**, 5670 (1991).
3) T. Inoue, E.-J. Hwang, and K. Osaki, *J. Rheology*, **36**, 1737 (1992).
4) K. Osaki, H. Okamoto, T. Inoue, and E.-J. Hwang, *Macromolecules*, **28**, 3625 (1995).
5) T. Inoue and K. Osaki, *Macromolecules*, **29**, 1595 (1996).
6) T. Inoue, H. Okamoto, Y. Mizukami, H. Matsui, and K. Osaki, *Macromolecules*, **29**, 6240 (1996).
7) T. Inoue, H. Matsui, S. Murakami, S. Kojiya, and K. Osaki, *Polymer*, **38**, 1215 (1997).
8) T. Inoue and K. Osaki, *Polym. J.* (*Tokyo*), **28**, 76 (1996).
9) T. Inoue, D.-S. Ryu, K. Osaki, and T. Takebe, *J. Polym. Sci. Polym. Phys. Ed*, **37**, 000 (1999).
10) T. Inoue, D.-S. Ryu, and K. Osaki, *Macromolecules*, **31**, 6977(1998).
11) T. Inoue, S. Onogi, and K. Osaki, submitted to *Macromolecules*.

Polymer-solvent organization in relation to chain microstructure

Jean-Michel GUENET

Université Louis Pasteur-CNRS UMR 7506
4, rue Blaise Pascal
F-67070 STRASBOURG Cedex FRANCE
guenet@ldfc.u-strasbg.fr

Abstract. Most polymers possess a microstructure (helicity, special sites bearing fractional charge arising from bonds polarization..) that has dramatic consequences on the manner they interact with solvent molecules to form organized systems. Here are reviewed some experimental results obtained on different varieties of *polystyrenes* (isotactic, syndiotactic), on *atactic polyvinyl chloride* and on *chemically-modified polyvinyl chloride* (chlorine atoms replaced either by hydrogen atoms or by thiophenate groups) that cast some light on these effects.

Introduction

Polymers in solution are usually portrayed as structureless random threads imbedded in spherical solvent molecules. Polymer-solvent interactions are accordingly accounted for with one single parameter, the χ_1 interaction parameter of the Flory-Huggins theory[1]. When polymer organization, such as crystallization, takes place from these solutions, then a simple relation can be established[1] between the melting point depression ($\Delta T = T_m^o - T_m$ between the melting point of the pure polymer and the polymer in presence of solvent), the polymer volume fraction φ_p and χ_1:

$$\Delta T / T_m T_m^o = \left[(1-\varphi_p) - \chi_1 (1-\varphi_p)^2\right] \qquad \textbf{(1)}$$

This relation implies that the better the interaction the larger the melting point depression. As will be discussed in this paper this relation is not always valid because most polymers possess a microstructure that has dramatic consequences on the manner they interact with solvent molecules and they form organized systems[2,3]. The chain microstructure may be defined as either the « *embossment* » created either by the heavy lateral groups or by the helical form taken on by the polymer, or/and the existence of well-defined sites bearing fractional charge arising from bonds polarization[4]. Similarly, solvent molecules are far from being sphere-like so that some sort of recognition mechanism will take place between polymer microstructure and solvent shape. This will strongly influence the molecular structure, the morphology and the mechanical properties of the organized systems. In this paper two cases will be briefly described: *polystyrenes* (isotactic and syndiotactic) and *poly[vynil chloride]*.

Polystyrenes gels and spherulitic assemblies

Two types of morphology shall be distinguished: *fibrillar morphology* to which most thermoreversible gels pertain, and *spherulitic morphology* arising from the growth of lamellar crystals, also termed chain-folded crystals for obvious reasons[3].

Depending upon the tacticity and the solvent different morphologies can be produced. For instance isotactic polystyrene forms fibrillar gels in decalin (*cis* or *trans*)[3,5] while syndiotactic polystyrene produces only spherulitic assemblies[6,7]. The reason why relies upon the stabilisation of the helical conformation in the case of isotactic polystyrene. Under these conditions, chain folding is impeded and so is the growth of spherulitic structures. The stabilisation process is highlighted by neutron scattering experiments [8](figure 2) that show two important results: i) the chain is worm-like in the *gel state* with a persistence length of about 4 nm, i.e. 4 times larger than that in the usual *unperturbed state*, ii) this conformation is kept in the *sol state* which hints at the existence of a stabilisation mechanism. On the other hand, it is known that polymer-solvent compounds are formed between iPS and both conformers of decalin[9]. Accordingly, Guenet and coworkers have suggested that solvent molecules are housed between the cavities designed by the phenyl groups of iPS chains when under a near-3_1 helical conformation which results in impeding chain folding[3].

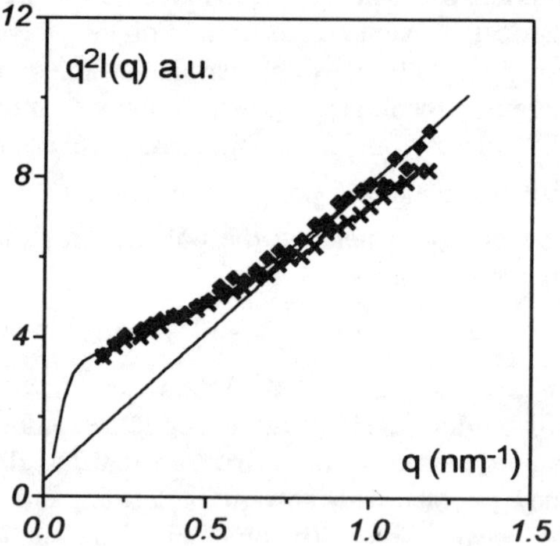

Fig. 1. Kratky-plot ($q^2I(q)$ vs q where $q= 4\pi/\lambda \sin \theta/2$; λ= neutron wavelength, θ= scattering angle) for a few labeled chains imbedded in a hydrogenous matrix. ✖= gel state, ◆= molten state at T= 66°C. The straight line stands for the 1/q asymptotic regime. The solid line is obtained by using Yoshisaki and Yamakawa semi-analytical expression [??].

The degree of solvation in the compound is highly dependent upon the solvent type: stoichiometries ranging from 0.6 solvent molecule/monomer (1-

chlorodecane) up to 1.75 solvent molecule/monomer (*cis*-decalin) have been reported[9,10]. Stoichiometries in *cis*-decalin and *trans*-decalin also differ markedly (1.75 for the former against 1.15 for the latter). Interestingly, in the case of 1-chlorodecane, for which there is less than 1 solvent molecule per monomer, the compound is not stable upon heating and transforms spontaneously into the non-solvated crystalline form (incongruently-melting compound)[10]. The notion of polymer-solvent compound implies the existence of a biphasic system: the *polymer-rich phase*, namely the compound, and the *polymer-poor phase* (in most cases almost pure solvent). The existence of polymer-solvent compounds, that are frequently designated as intercalate compounds or crystallosolvates, has a direct bearing upon the mechanical properties, more specifically upon the relation between modulus and polymer concentration.

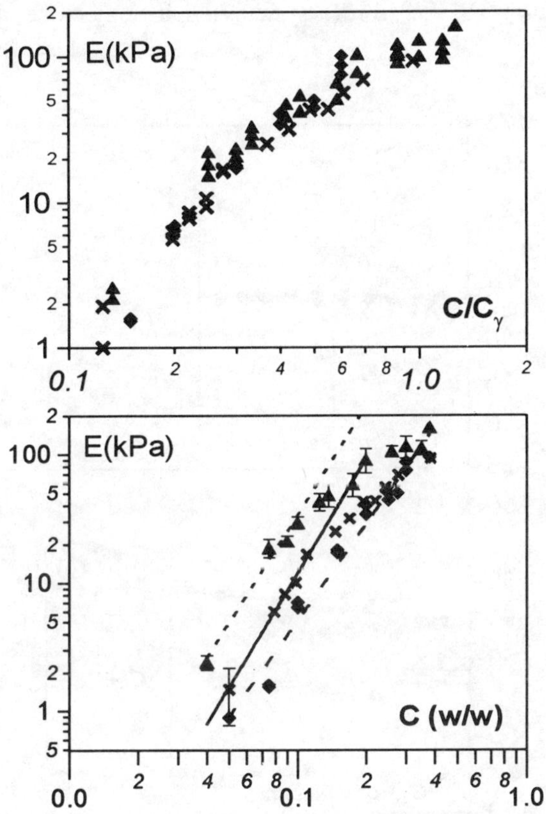

Fig. 2. *lower figure*: elastic modulus vs polymer concentration for gels prepared with iPS in cis-decalin (▲), in trans-decalin (✖), and 1-chlorodecane (◆). *upper figure*: the same data plotted now as a function of the polymer-rich fraction instead.

As can be seen in figure 2 (lower), there is a large discrepancy between the values of the elastic modulus found for isotactic polystyrene gels in *cis*-decalin, *trans*-decalin and 1-chlorodecane[11]. If equation (1) applied, there should not be any noticeable difference as the Θ-temperature for the three solvent is close enough. Here, the existence of polymer-solvent compounds of differing stoichiometries must be taken into account. ***The relevant parameter*** is no longer the polymer volume fraction ***but the volume fraction of the phase responsible for the elastic properties***. ***This phase is nothing but the polymer-rich phase***. At the same polymer volume fraction, the higher the stoichiometry, i.e. the lower the stoichiometric concentration, the higher the fraction of the polymer-rich phase. Provided that the polymer concentration of the polymer-poor phase is low enough, then the fraction of polymer-rich phase is simply derived by rescaling the polymer volume fraction by the polymer volume fraction at the stoichiometric conditions ($\varphi/\varphi_\gamma \sim C/C_\gamma$). Plotting the elastic modulus E vs C/C_γ instead of C produces an universal behaviour (figure 2, upper), which points at a similar mesoscopic morphology for all the three systems that otherwise possess differing stoichiometries.

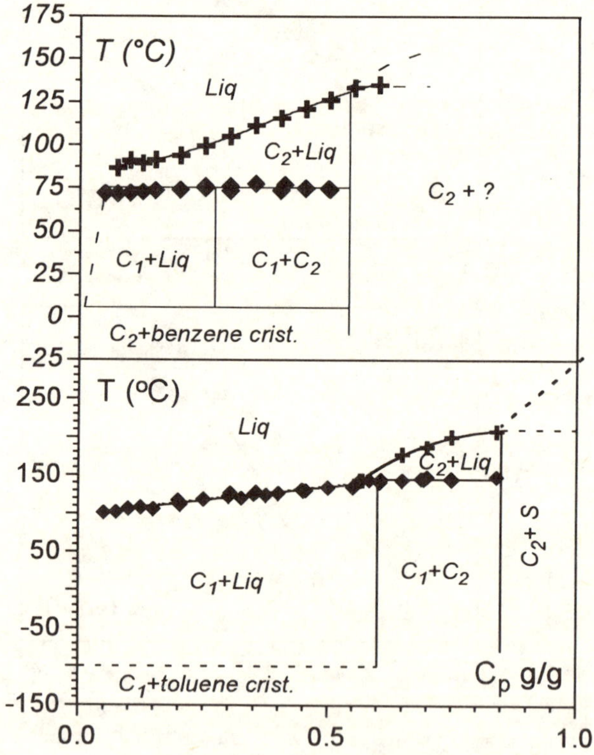

Fig. 3. Temperature-Concentration phase diagrams for sPS/benzene (*upper*) and sPS/toluene (*lower*). C= compound, S= solid phase. Upper: C_1= *4 benzene/monomer*, C_2= *1 benzene/monomer*. In toluene C_1= *0.8 toluene/monomer*, C_2= *0.25 toluene/monomer* (The latter compound was already identidied by Chatani's et al. on solvent-exposed samples.

The effect of the solvent shape is also conspicuous with syndiotactic polystyrene in the case of *benzene* and *toluene*. The temperature-concentration phase diagrams reveal the occurrence of compounds of differing stoichiometries. As can be seen a highly-solvated compound is formed in *benzene* (about *4 benzene molecules/monomer*) whereas the stoichiometry of the most solvated compound compound in *toluene* is only *0.8 toluene/monomer*[12,13]. The solvent microstructure definitely alters the degree of solvation of the compounds in spite of a virtually identical Flory's interaction parameter for both solvents. Another interesting outcome is given by comparison of iPS and sPS systems in *trans*-decalin. While compounds occur for both polymers in this solvent with about the same stoichiometry [14,15], a fibrillar gel forms with iPS whereas spherulites (chain-folded crystals) grow in the case of sPS.

These two items highlight the subtlety of the stabilization mechanism of the helicoidal form in relation to the chain microstructure. The cavity in sPS arising from the 2_1 helical form is slightly more spacious than a *trans*-decalin molecules, hence a poor stabilisation effect[16]. Conversely, the cavity created by the phenyl rings of the iPS 3_1 helix possesses the adequate space for housing a *trans*-decalin molecule, hence a better stabilisation.

Gels from Poly[vinyl chloride] and from chemically-modified Poly[vinyl chloride]

Poly[vinyl chloride] gels.

This polymer possesses a very high propensity to produce thermoreversible gels in a large variety of organic solvents [17]. While it now makes little doubt that this capability arises from the crystallization of the *highly-syndiotactic* sequences, the high values found for the elastic modulus suggest the occurrence of other physical links. As a matter of fact, the low fraction of syndiotactic sequences, including those containing some isotactic defects, does not exceed 10% [18], and cannot account for the magnitude of the elastic modulus, especially so in the light of the findings by He et al.[19]. These authors have synthesized a multiblock copolymer mimicking PVC as 10% of crystallizable sequences alternate with 90% of flexible sequences. They have observed that the modulus of a thermroeversible gel made up of this polymer is some 100-fold lower than that of PVC gels at the same polymer concentration. Admittedly, the existence of additional links must be seriously contemplated.

PVC gels possess a fibrillar morphology [20,21], and are therefore biphasic systems. Guenet and Mutin have further suggested on the basis of the phase diagram drawn in figure 4 that there occurs a polymer-solvent compound in these gels [21]. Indeed, this diagram displays a non-variant transition at 50°C which implies the existence of a *peritectic transformation*. Such a transformation usually takes place for non-stoichiometric compounds.

As the *highly-syndiotactic* sequences are known to crystallize without solvent occluded, this compound can only form with the less stereoregular sequences.

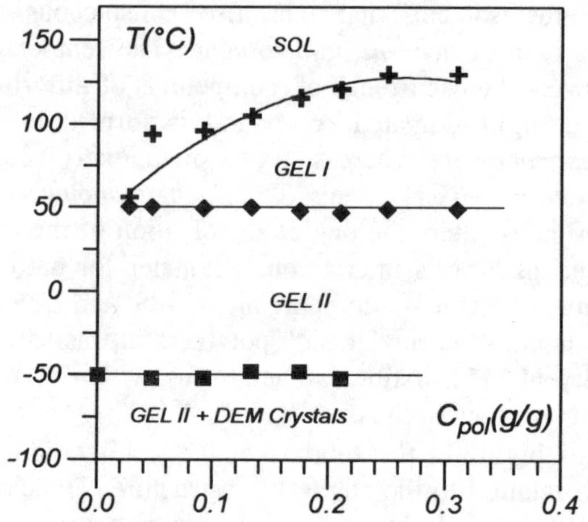

Fig. 4. Temperature-concentration phase diagram for PVC/diethyl malonate gels. The transition from GEL II to GEL I is accompanied by a sharp drop of the elastic modulus.

Mutin and Guenet, and later Abied et al. [22] have devised a molecular mechanism based on electrostatic interactions to explain the formation of this complex (see figure 5). It is known from spectroscopic investigations that the Cl-C-H bond is polarized which entails the appearance of fractional electric charges[23]. The same phenomenon occurs on diesters that can then interact simultaneously with two different chains. Inferring that this process can be cooperative, then PVC gel fibrils can be strengthened but also additional links

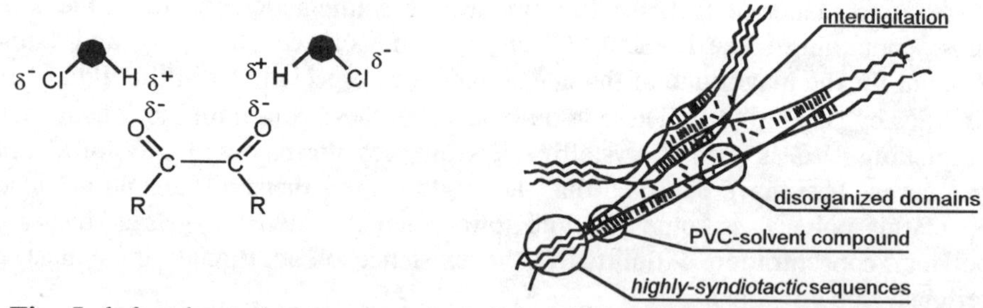

Fig. 5. *left*: schematic representation of the elctrostatic interaction of a diester molecule with two PVC chains (seen parrallel to their axis). *Right*: molecular model for the PVC fibrils and the way they can form additional physical links by an interdigitation process.

between fibrils can be created by some interdigitation process [24] as portrayed in figure 5. The occurrence of such a compound with the molecular structure propose by Abied et al.[22] would solve the question relative to the high modulus values.

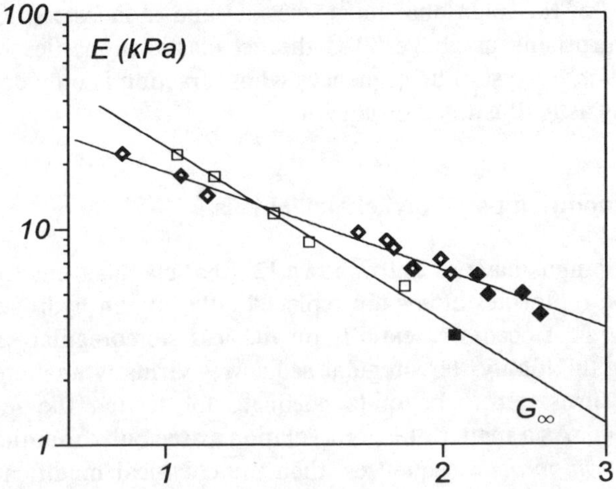

Fig. 6. Elastic modulus vs degree of swelling (v/v) for a series of diethyl esters (◇), dibutyl esters (◆), and ethyl esters (◘). ■= isoamy acetate.

On the basis of the above molecular model, Najeh et al. [25] reasoned that diesters and monoesters should produce gels with differing characteristics as monoesters are unable to bridge chains together unlike diseters, although they could also form compounds. In order to overcome the problem of solvent quality, these authors compared several series of diesters and monoesters starting from the same PVC concentration and allowing the samples to swell to equilibrium in an excess of preparation solvent. Samples reaching the same degrees of swelling would accordingly correspond to solvents of same quality.

As is shown in figure 6, these authors then compared the elastic modulii of the samples prepared in the different esters. Diesters and monoesters do behave differently. Diesters fall on the same line while monoesters display another behaviour. One would have, however, expected the modulus of gels prepared with diesters to be systematically higher than their monoester counterparts. This is no so because another effect takes place: the disordered domains in monoesters stand below their glass transition temperature[25], which also entails an increase of the fibrils intrinsic modulus. By increasing the solvent quality of monoesters toward PVC, the modulus drops down rapidly due to a platicization phenomenon. This is the reason why the variations differ whether monoester or diesters are considered. In diesters, the formation of a « bridging » compound does not depend upon solvent quality. As suggested by Najeh et al. the decrease of modulus when increasing diester solvent quality can be interpreted by assuming a simple « dilution » of the number of interfibrillar junctions per unit volume.

The difference observed above is also highlighted by viscometry experiments on aggregates produced below the critical gel concentration. The behaviour as a function of the preparation concentration differs whether diesters or monoesters are used as long as temperature stands below 50°C[26]. Above this temperature, which corresponds to compound melting, and correspondingly to the disappearance of the additional links, diesters and monoesters behave the same. This is not surprising as above 50°C the gel elastic properties rely only on the *highly-syndiotactic* crystalline sequences whose fraction is only determined by the solvent quality as is illustrated by equation 1.

Chemically-modified poly[vinyl chloride] gels.

Lopez and Mijangos have recently shown [27]that chemical modification of PVC by substitution (chlorines atoms are replaced either by an hydrogen atom or by a thiophenate group) occurs essentially on the less stereoregular sequences of the chain, leaving the highly-stereoregular sequences virtually unaltered. This type of modification turns out to be quite adequate for testing the molecular model described above. As a matter of fact, if gelation arises only from the crystallization of the *highly-syndiotactic* sequences, then the chemical modification should not affect at all the gel mechanical properties. As shown in figure 7, Lopez and Mijangos have observed that a significant decrease in gel modulus is observed for low degrees of modification when replacing Cl by H. The effect is more dramatic when Cl is substituted by a thiophenate group. These experiments provide a strong support to the molecular model devised by Abied et al..

What occurs exactly at the molecular level is unveiled by small-angle neutron scattering experiments on gels and light scattering experiments on aggregates. SANS experiments are consistently interpreted with fibrils wherein solvent is

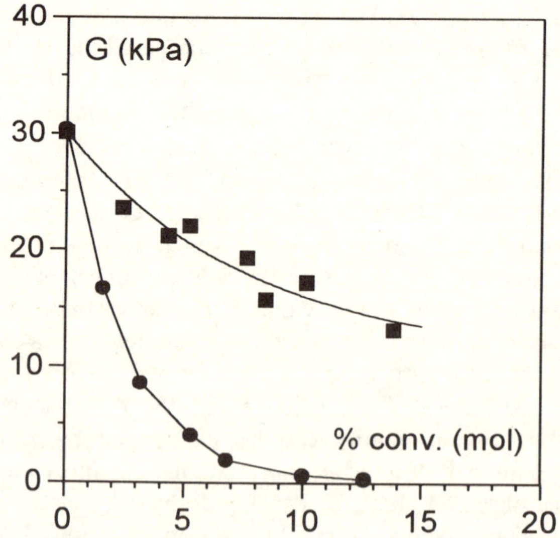

Fig. 7. Variation of the elastic modulus as a function of the degree of conversion for gels prepared in diethyl oxalate. (■) Cl replaced by H; (✚) Cl replaced by a thiophenate group. After Lopez and Mijangos [27].

occluded. The scattering function can be accounted for by considering cross-section polydispersed fibrils characterized by a distribution function of the type $w(r) \sim r^{-1}$ bounded by two radii r_{min} and r_{max}. The outcome of chemical modification has little effect upon the morphology of the basic fibrils but the scattering curves indicate a drop of the number of efficient contacts between fibrils through physical links. The light scattering experiments gives additional information. Results on non-modified samples are consistent with fibrils of longitudinal fractal dimension $D_f = 1.5$ [26] as is illustrated by the variation of the fibril radius of gyration with molecular weight ($R_G \sim M^{0.66}$) The value of this fractal dimension implies that gel fibrils are not straight objects. Incidentally, this fractal dimension allows one to account for the variation of the elastic modulus with PVC concentration[26]. In the case of chlorine atoms replaced by hydrogen atoms, the same variation is found, yet the fibrils' molecular weight increases by about 2 (degree of modification 6%) while the radius of gyration remains virtually a constant[29]. According to Reinecke et al. [29]this result suggests that fibrils' mean cross-section increases while their mean length remains unchanged. Conversely, when Cl is replaced by a thiophenate group, both M and R_G decrease which implies that the length of the fibrils is considerably modified. These results and interpretations are altogether qualitatively consistent with the behaviour of the modulus as a function of the degree of modification. Indeed, the slight increase of fibrils' cross-section (about 40%) entails only a moderate decrease of the number of junctions between fibrils. Conversely, a strong decrease of the fibrils' length will create dangling fibrils that no longer participate in the gel elasticity.

Concluding remarks

The two cases presented here highlight the major role played by the chain microstruture in the growth, the morphology and the properties (elastic property for instance) of organized systems obtained from polymer solutions. So far there is no convenient parameter(s) to derive a theoretical approach on this issue. The most convenient procedure way to make some predictions on the behaviour of such polymer solutions remains the use of computer molecular modelling. At any rate, the first thing to be established ought to be the experimental temperature-concentration phase diagram. Knowledge of the phase diagram is essential because the behaviour of these systems depend upon solvent shape (type, conformers,..), temperature and concentration[30]. This type of diagram provides a wealth of useful information that should guide further investigations and should prevent useless controversies.

References

1) Flory, P.J. *Principle of Polymer Chemistry* **1953** Cornell Univ. Press, Ithaca, NY
2) Point, J.J.; Coutelier, C.; Vilers, D. *J. Polym. Sci. Polym. Phys. Ed.* **1985** *23* 231
3) Klein, M.; Mathis, A.; Menelle, A.; Guenet, J.M. *Macromolecules* **1990** *23* 4591
4) Monteiro, E.E.C.; Mano, E.B. *J. Polym. Sci. Polym. Phys. Ed.* **1984** *22* 533
5) Atkins, E.D.T.; Hill, M.J.; Jarvis, D.A.; Keller, A.; Sarhene, E.; Shapiro, J.S. *Colloid Polym. Sci.* **1984** *262* 22
6) Prasad, A.; Mandelkern, L. *Macromolecules* **1990** *23* 5041
7) Daniel, C.;Dammer, C.; Guenet, J.M. *Polymer* **1994** *35* 4243
8) Klein, M.; Brûket, A.; Boué, F.; Guenet, J.M. *Polymer* **1991** *32* 1943
9) Guenet, J.M.; McKenna, G.B. *Macromolecules* **1988** *21* 1752
10) Guenet, J.M.; Lotz, B.; Wittmann, J.C. *Macromolecules* **1985** *18* 420
11) McKenna, G.B.; Guenet, J.M. *J. Polym. Sci. Polym. Phys. Ed.* **1988** *26* 267
12) Daniel, C.; De Luca, M.D.; Brûlet, A.; Menelle, A.; Guenet, J.M. *Polymer* **1996** *37* 1273
13) Daniel, C.; Brûlet, A.; Menelle, A.; Guenet, J.M. *Polymer* **1997** *38* 4193
14) Roels, T.; Deberdt, F.; Berghmans, H. *Macromolecules* **1994** *27* 6216
15) Daniel, C. *Thesis* **1996** Université Louis Pasteur Strasbourg (F)
16) Immirzi, A.; de Candia, F.; Ianelli, P.; Zambelli, A.; Vittoria, V. *Makromol. Chem. Rapid Commun.* **1988** *9* 761
17) Guenet, J.M. in *Thermoreversible Gelation of Polymers and Biopolymers* **1992** Academic Press, London
18) Lemstra, P.J.; Keller, A.; Cudby, M. J. Polym. Sci. Polym. Phys. Ed. **1978** 16 1507
19) He, X.W.; Herz, J.; Guenet, J.M. *Macromolecules* **1988** *21* 1757
20) Yang, Y.C.; Geil, P.H. *J. Macromol. Sci.* **1983** *B(22)2* 463
21) Mutin, P.H.; Guenet, J.M. *Macromolecules* **1989** *22* 843
22) Abied, H.; Brûlet, A.; Guenet, J.M. *Colloid and Polym. Sci.* **1990** *268* 403
23) Tabb, D.L.; Konig, J.L. *Macromolecules* **1975** *8* 829
24) López, D.; Dahmani, M.; Mijangos, C.; Brûlet, A.; Guenet, J.M. *Macromolecules* **1994** *27* 7415
25) Najeh, M.; Munch, J.P.; Guenet, J.M. *Macromolecules* **1992** *25* 7018
26) Dahmani, M.; Fazel, N.; Munch, J.P.; Guenet, J.M. *Macromolecules* **1997** *30*
27) López, D.; Mijangos, C.; Muñoz, M.E.; Santamaria, A. *Macromolecules* **1996** *29* 7108
28) Dahmani, M.; Skouri, M.; Munch, J.P.; Guenet, J.M. *Europhysics letter* **1994** *26* 19
29) Reinecke, H.; Fazel, N.; Dosière, M.; Guenet, J.M. *Macromolecules* **1997** *30*
30) Guenet, J.M. *Thermochim. Acta.* **1996** *284* 67

author's present address:
Institut de Chimie des Surfaces et Interfaces
CNRS UPR 9069
BP 2488
15, rue J. Starcky
F-68057 MULHOUSE Cedex FRANCE

Polymer-surfactant interaction in thermoreversible gels

Fumihiko TANAKA

Department of Polymer Chemistry, Graduate School of Engineering,
Kyoto University, Kyoto 606-8501, Japan
ftanaka@kuic.kyoto-u.ac.jp

Abstract

The influence of low molecular weight surfactants on the sol/gel transition of hydrophobically modified water-soluble associating polymers is studied on the basis of the recent theory of thermoreversible gelation with multiple junctions. It is shown that the existence of a lower and an upper bound in the aggregation number of hydrophobes at the network junctions leads to non-monotonic polymer gelation concentration as a function of the surfactant concentration. There is a certain surfactant concentration where gelation is most promoted, thereby exhibiting a peak in the viscosity and modulus. Relation between this peak concentration and the critical micelle concentration of the added surfactant is detailed. The number of elastically active network chains is calculated in the post-gel regime as a function of the surfactant concentration, and compared with the experimental data on the high frequency storage modulus of HEUR/SDS systems.

Introduction

The interaction between polymers and surfactants have been a subject of great interest. The problem was laid initially in studies of proteins associated with natural lipids, and later in the studies of their association with synthetic surfactants. More recently, interaction of water-soluble synthetic polymers such as poly(ethylene oxide) with ionic and non-ionic surfactants[1, 2, 3, 4] have attracted interest of researchers because of the scientific and technological implications. Adding surfactants to polymer solutions with formation of polymer/surfactant complex can substantially alter the physical properties of the starting polymers. The effect can be summarized into the following four categories: (i) Conformational transition of polymers such as coil-globule transition and coil-rod transition, (ii) Expansion and shift of the phase separation region on the polymer/solvent phase plane, (iii) Formation of composite microphases, (iv) Shift of the sol/gel transition line and modification of the rheological properties.

When polymers carry small fraction of hydrophobic groups, effects are dramatically enhanced as is seen in some of the above examples. Such polymers are often referred to as "associating polymers" because they show tendency of self-assembly, leading to gelation in the extreme cases, caused by the aggregation of hydrophobes. The ability of surfactant binding is enhanced by the hydrophobic interaction between polymer hydrophobes and surfactant hydrophobes. Thus, typical model polymers such as hydrophobically ethoxylated

Y. Morishima · T. Norisuye · K. Tashiro (Eds.)
Molecular Interactions and Time-Space Organization in Macromolecular Systems
Proceedings of the OUMS '98, Osaka, Japan, 3-6 June, 1998

urethane (HEUR) and ethyl hydroxyethyl cellulose (EHEC), hydroxypropyl methyl cellulose (HPMC) have recently been the focus of the study. Profound influence of added surfactants on the rheological properties has been reported. For example, the high-frequency plateau modulus of HEUR solution exhibits a peak when sodium dodecyl sulfate (SDS) is added at low polymer concentrations. The peak disappears at higher polymer concentrations[5].

On the basis of our recent theory of thermoreversible gelation with multiple junctions[6], we present in this paper a theoretical description of sol/gel transition in the polymer/surfactant mixtures.

Mixture of associating polymers and surfactants

We consider a model mixture of associating polymers and low molecular-weight surfactant molecules in a solvent. Each polymer is assumed to carry the number $f(\geq 2)$ of associative groups (hydrophobes in the case of HEUR) along its chain, and each surfactant molecule is modeled as a low molecular weight molecule carrying a single hydrophobe connected to the hydrophilic head. This is a special case of the model solution treated in our previous study[6] (referred to as TS) on the associating polymers capable of forming networks with multiple cross-link junctions. We therefore start from the lattice-theoretical free energy detailed in it.

In equilibrium, hydrophobes on the polymers and on the surfactants aggregate into micelles that work as cross-link junctions in finite-size clusters, and also in networks at high enough concentrations. The aggregation number differs from one micelle to another. To specify the size of the junctions, we introduced a concept of the *junction multiplicity* in the previous study TS. The multiplicity k is defined by the number of chains combined together into a junction. It agrees with the aggregation number of the micelles in the case of associative hydrophobes. Thus $k = 1$ indicates unassociated hydrophobes. Let p_k be the probability for a randomly chosen hydrophobe to belong to a junction of the multiplicity k at a certain given temperature, polymer- and surfactant concentration. Then $p_1 \equiv 1 - \alpha$ is the probability for a hydrophobe to remain unassociated, where α is the extent of reaction (or conversion) in the conventional meaning. It was shown in TS that, from the normalization condition $\sum p_k = 1$, the total concentration ψ of the hydrophobes should satisfy the relation

$$\lambda(T)\psi = zu(z), \tag{1}$$

where $\lambda(T)$ is the association constant, and the function $u(z)$ to be used to characterize micellar junctions is defined by

$$u(z) \equiv \sum_{k=1}^{\infty} \gamma_k z^{k-1}, \tag{2}$$

where the coefficient γ_k comes from the surface free energy of the micelle of aggregation number k. The parameter z that appeared in this relation is defined by $z \equiv \lambda(T)\psi p_1 = \lambda(T)\psi(1-\alpha)$ and gives the (reduced) concentration of the hydrophobes that remain unassociated in the solution. If we use the (reduced) volume fraction $c_f \equiv \lambda(T) f \phi_f / n_f$ instead of ψ_f for polymers and

$c_1 \equiv \lambda(T)\phi_1/n_1$ for surfactants, (n_i being the number of elementary statistical units on the molecule of species i) the relation (1) can be transformed into

$$c_f(1+\eta) = zu(z), \tag{3}$$

where $\eta \equiv c_1/c_f$ is the ratio of the surfactant concentration to the polymer concentration. Solving this relation with respect to z, we find z, and hence the conversion α as a function of a given temperature and concentration.

As the concentration is increased, the weight-average cluster size increases, and eventually becomes infinite at the sol-to-gel transition point. For our polymer/surfactant system, this sol/gel transition condition is explicitly given by

$$(f-1)zu'(z)/(1+\eta)u(z) = 1. \tag{4}$$

Combining this condition with the relation (3) and eliminating the parameter z, we find the sol/gel transition curve on the temperature-concentration plane.

To see how the sol/gel transition concentration shifts by adding surfactant molecules, we have to specify the multiplicity of the junctions in more detail. At this stage we have to stress the importance of both the lower and upper limit in the multiplicity allowed in order to describe the non-monotonic behavior of the observed modulus and viscosity. We therefore introduce a model function

$$u(z) = 1 + \sum_{k=s_{min}}^{s_{max}} z^{k-1} = 1 + (z^{s_{min}-1} - z^{s_{max}})/(1-z), \tag{5}$$

where s_{min} is the minimum multiplicity and s_{max} the maximum multiplicity allowed. We have neglected possible contribution to the free energy from micellar surface and set all $\gamma_k = 1$ for $s_{min} \leq k \leq s_{max}$. When polymer concentration is low and the number of hydrophobes is not enough to form junctions, addition of surfactants combines the unassociated hydrophobes until their aggregation number exceeds s_{min} and stabilize them. In this situation, the surfactant works as a cross-linking agency. On the contrary, when the polymer concentration is large and many junctions are already formed, some of the polymer hydrophobes in the junctions are replaced by surfactant hydrophobes, and lead to the dissociation of network junctions.

Fig. 1 shows how junctions are formed and destroyed by added surfactants in the special case where the multiplicity is fixewd at $s_{min} = s_{max} = 5$. From these considerations, we expect that there is no surfactant-mediated process if no minimum multiplicity exists, *i.e.*, if $s_{min} = 2$. In such a special case, hydrophobes form stable junctions no matter how small their aggregation number may be. The addition of surfactants therefore simply destroys the already existing junctions.

To demonstrate these ideas, we calculate, by solving (3) and (4), the concentration of polymers at the sol/gel transition point as a function of the concentration of the added surfactant. Fig. 2 shows the result for the telechelic ($f = 2$) polymers. Both polymer and surfactant concentrations are expressed in terms of the reduced concentration, the number of hydrophobes (per lattice cell) times association constant. To see the effect of the minimum multiplicity,

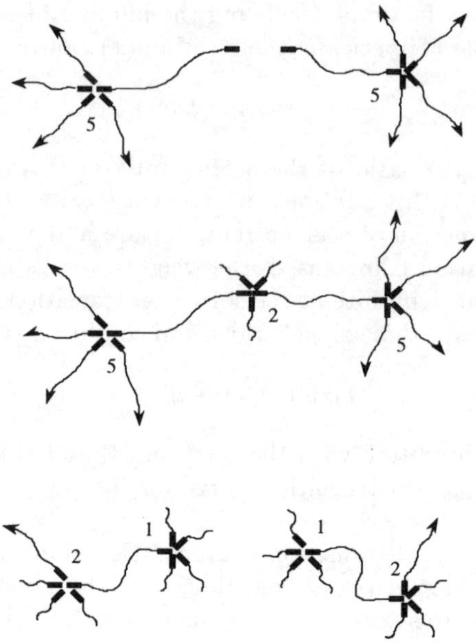

Figure 1: Formation of a junction with the help of surfactant molecules (surfactant-mediated association) and destruction of a junction by excess surfactant molecules. The allowed multiplicity is fixed at $s = 5$. Figures near the junctions indicate their branching numbers. Average branching number is (5+5)/2=5 (top figure), (5+5+2)/3=4 (middle figure), (2+2+1+1)/4=1.5 (bottom figure). It monotonically decreases with surfactant concentration.

s_{min} is varied from curve to curve, while the maximum multiplicity is fixed at $s_{max} = 8$. It is clear that the sol/gel concentration c_f^* monotonically increases with the surfactant concentration for $s_{min} = 2$ (no lower bound), *i.e.*, gelation is blocked by the surfactant. But if there is a gap between $k = 1$ (unassociated) and $k = s_{min}$, a minimum in c_f^* starts to appear. At this surfactant concentration gelation is most promoted as can be seen for $s_{min} \geq 3$. The surfactant concentration at which c_f^* becomes minimum (hereafter referred to as *surfactant-mediated gelation* point SMG) increases as the gap becomes larger.

CMC of the surfactant molecules

It is known that surfactant molecules form micelles above a certain concentration. The concentration at which micelles start to appear is referred to as critical micelle concentration (CMC). To see how CMC is affected by the presence of associating polymers, let us consider the surfactant molecules that are not associated to any polymers.

We define the critical micelle concentration (CMC) by the surfactant concentration at which micelles consisting of only surfactant molecules start to appear. In what follows, therefore, we refer to this concentration as *critical*

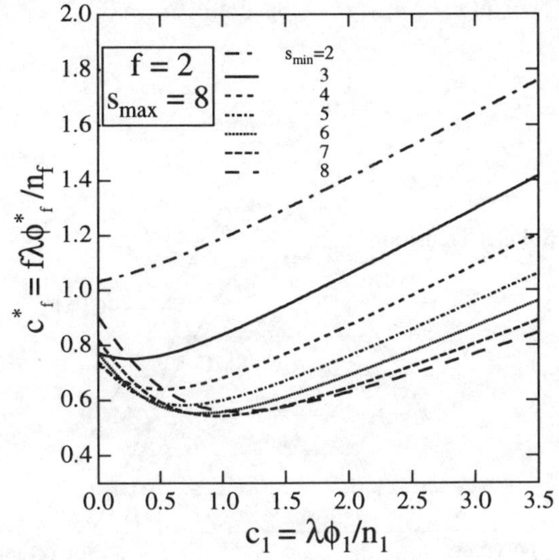

Figure 2: Polymer concentration at sol/gel transition as a function of the concentration of added surfactant. The concentration is measured in terms of the number density $f\phi/n$ of the hydrophobes times the association constant $\lambda(T)$ for each species. Minimum multiplicity s_{min} is varied from curve to curve under a fixed maximum multiplicity s_{max}. While the sol/gel concentration monotonically increases with surfactant concentration for $s_{min} = 2$ (*i.e.*, there is no lower bound in the multiplicity), there appears a minimum at a certain surfactant concentration for $s_{min} \geq 3$ (*i.e.*, if there is a gap below the minimum allowed multiplicity). In the latter case, gelation is promoted by the surfactant molecules, and referred to as surfactant-mediated gelation (SMG) in the text.

pure micelle concentration (CPMC). One conventional criterion for CMC is to find the concentration at which the osmotic pressure changes its slope most rapidly[7]. This criterion is almost equivalent to the condition that the weight density of the surfactant molecules expressed as a function of that of the isolated (unassociated) molecules ceases to have its inverse function[7]. In our present particular model, this happens when

$$d(xu(x))/dx = u(x) + xu'(x) = 0 \tag{6}$$

holds. This equation is an algebraic equation for x and has its roots on the complex x-plane. These roots are branch points of the inverse function. When the concentration x_1 of the unassociated surfactant passes near the root that lies closest to the real x-axis, the osmotic pressure due to surfactant molecules changes its slope most rapidly.

To study the relative position of CPMC and SMG concentration, we consider a special multiplicity model in which junction multiplicity is fixed at a single value $s_{min} = s_{max} \equiv s$. The function $u(x)$ in this fixed multiplicity model takes the form $u(x) = 1 + x^{s-1}$ by definition.

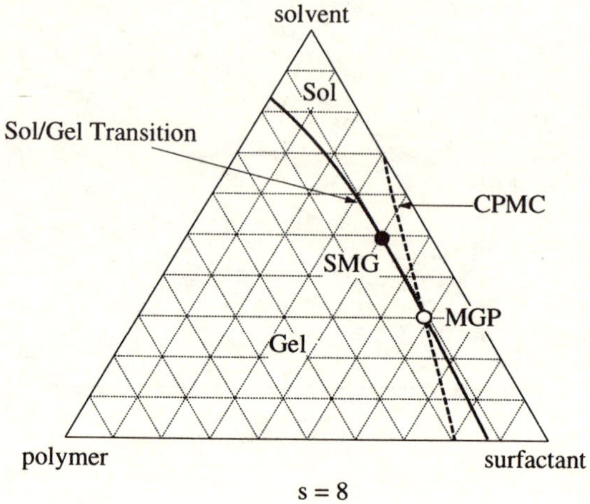

Figure 3: The sol/gel transition line and the CPMC line drawn on the ternary phase plane of polymer/surfactant/solvent system. The minimum gelation point is indicated by a black circle. The white circle shows a special point at which gelation and micellization simultaneously take place.

Fig. 3 shows sol/gel transition line (solid line) and CPMC line (broken line) on the the ternary phase plane of polymer/surfactant/water system for $s = 8$. There is an intersection between solid and broken line for each given multiplicity. This is a special point where sol/gel transition and CPMC take place simultaneously. We call this special point *micellization gelation point* (MGP). For instance this point for $s = 8$ is located at the surfactant concentration nearly twice as large as the SMG concentration. SMG and MGP are indicated by the black and white circle. Their relative position may change if we allow difference in the binding free energy for polymer hydrophobe and for surfactant hydrophobe.

High frequency elastic modulus

Let us proceed to the study of the dynamic mechanical moduli. In the experiment of HEUR/SDS system[5], addition of surfactant results in several effects. The moduli are no longer described by the simple Maxwell element with a single relaxation time, but a shoulder appears on the loss modulus at higher frequencies. Furthermore, the high frequency plateau in the storage modulus reveals non-monotonic dependence on the SDS concentration. At low polymer concentration, it initially rises to a peak and then decreases monotonically, falling eventually to zero at higher SDS concentration. With increase in the polymer concentration, the height of the peak decreases and its position shifts to lower SDS concentration. Above a certain polymer concentration, the peak disappears. Finally, the average rheological relaxation time also shows a peak for all polymer concentrations measured.

To see these observations from a theoretical point of view, let us first count the number of polymer chains in the network that contribute to the mechanical moduli. A chain in the network can be classified into two main categories: an elastically effective chain and a dangling chain. An elastically effective chain is a chain whose both ends are connected to the different network junctions. A dangling chain is a chain that carries a group dangling from the network. Stress is transmitted only through the elastically effective chains. The criterion to judge whether a chain is effective or ineffective is, however, not unique. Several candidates have been examined in the theoretical review on rubber elasticity. Here, we employ the most precise criterion proposed by Scanlan[8] and Case[9]. The Scanlan-Case criterion assumes that those chains whose both ends are connected to the junctions that carry at least three paths leading to the network matrix are elastically effective. A junction with one path to the gel unites a group of subchains dangling from the network matrix whose conformations are not affected by an applied stress. A junction with two paths to the gel merely extends the length of an effective subchain.

To see this in more detail, let us specify the junction type as in [10] (referred to as TI). A junction of multiplicity k that is connected to the gel network through i paths is referred to as an (i,k)-junction. Let $\mu_{i,k}$ be the number of junctions specified by the type (i,k) for $k = 1,2,3,4,\cdots$ and for $0 \leq i \leq 2k$. We now employ the aforementioned criterion of Scanlan-Case[8, 9] to decide whether a subchain connected to two junctions of the type (i,k) and (i',k') at its ends is elastically effective or not. The criterion says that only subchains connected at both ends to junctions with at least three paths to the gel are elastically effective. We thus have $i, i' \geq 3$ for an effective chain. We may call the junctions with $i > 3$ *elastically effective junctions*. An effective subchain is then defined as a chain that is connected to two effective junctions at its both ends, and the total number of effective chains is given by

$$\nu_{eff} = \frac{1}{2}\sum_{k=2}^{\infty}\sum_{i=3}^{2k} i\mu_{i,k} \tag{7}$$

Fig. 4 shows the number $\nu_{eff}(c_1)$ of elastically effective chains plotted against the surfactant concentration c_1 (devided by the polymer concentration c). The number is normalized by the value $\nu_{eff}(0)$ under the absence of the surfactant. This ratio therefore gives the relative strength $G_\infty(c_1)/G_\infty(0)$ of the high frequency plateau value in the storage modulus. Polymers are assumed to carry two functional groups ($f = 2$). The allowed multiplicity of a junction ranges from 3 to 8. The polymer concentration is changed from curve to curve. As expected, the curves for low polymer concentrations first rise to a peak and then monotonically decreases to zero where gel network is broken into sol by the surfactant. For larger polymer concentrations, however, the curves do not show any peak, because junctions are well developed without surfactant molecules for these polymer concentrations, and the added surfactant merely destroys the junctions.

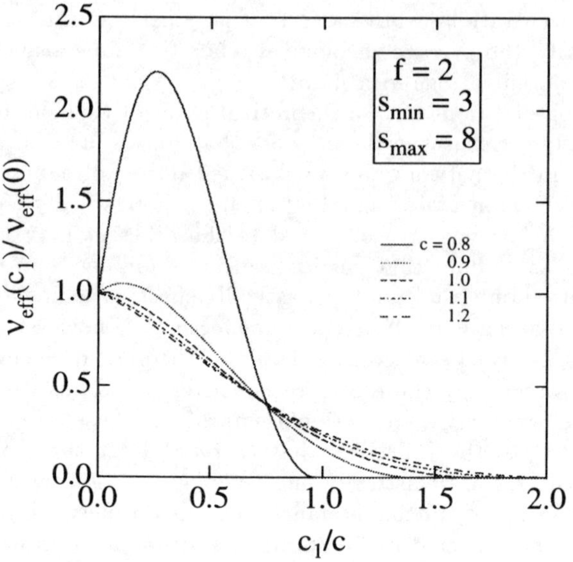

Figure 4: The number of elastically effective chains plotted against the surfactant concentration. For low polymer concentrations, it shows a peak at a certain surfactant concentration.

These calculations reproduce, at least qualitatively, the experimental data on such as HEUR/SDS reported by Annable *et.al.* (Figure 3 in the reference 5). In our theory all curves cross each other at a certain surfactant concentration, whereas the experimental data reveal the same tendency only for relatively higher polymer concentrations. The maximum in the modulus is caused by the existence of a forbidden gap in the multiplicity of the network junctions.

Conclusion and discussions

We have studied interaction of associating polymers with surfactants, thereby stressing how sol/gel transition is shifted and how the structure and mechanical properties of the network are modified. It was shown that observed non-monotonic change in rheological behavior with added surfactants is explained by the existence of a lower- and an upper bound in the multiplicity of junctions.

We also stressed that there is a special point on the polymer/surfactant/water ternary phase plane where gelation and critical (pure) micelle concentration simultaneously take place. In the preceeding study of TS, we showed that the osmotic compressibility reveals a singularity across the sol/gel transition line due to the loss of the translational entropy of the macroscopic cluster. Gelation can therefore be regarded as a thermodynamic phase transition in Ehrenfest's sense. On the other hand, derivative of the osmotic pressure abruptly falls at CMC although it shows no true analytical singularity. This confluent point is therefore to some extent similar to a higher order critical point where two different phase transition lines cross each other. Implication of new critical phenomena around this point is to be studied.

References

[1] B.J. Cabane, *J. Phys. Chem.* **81**, 1639 (1977).

[2] W. Brown,J. Fundin, M. da Graça Miguel, *Macromolecules* **25**, 7192 (1992).

[3] E. Feitosa, W. Brown, P. Hansson, *Macromolecules* **29**, 2169 (2169).

[4] E. Feitosa, W. Brown, M. Vasilescu, M. Swason-Vethamuthu, *Macromolecules* **29**, 6837 (6837).

[5] T. Annable, R. Buscall, R.Ettelaie, P. Shepherd, D. Whittlestone, *Langmuir* **10**, 1060 (1994).

[6] F. Tanaka, W.H.Stockmayer, *Macromolecules* **27**, 3943 (1994).

[7] F.H. Stillinger, A. Ben-Naim, *J. Chem. Phys.* **74**, 2510 (1981).

[8] J. Scanlan, *J. Polym. Sci.* **43**, 501 (1960).

[9] L.C. Case, *J. Polym. Sci.* **45**, 397 (1960).

[10] F. Tanaka, M.Ishida, M., *Macromolecules* **29**, 7571 (1996).

Viscoelastic Phase Separation and Transient Formation of Spongelike Patterns

Hajime Tanaka

Institute of Industrial Science, University of Tokyo, Minato-ku, Tokyo 106, Japan.
Tanaka@iis.u-tokyo.ac.jp

Abstract. It has so far been believed that phase separation in isotropic condensed matter can be classified into either solid or fluid models, including some modification due to elastic effects. Contrary to this common belief, we recently found unusual phase separation, which can be described by neither of the above models, in a mixture whose components have a large difference in their dynamics. Thus, we proposed that a new model of phase separation, namely, "viscoelastic model", is necessary to describe phase separation in such dynamically asymmetric mixtures. This model is likely a general model that can describe all types of isotropic phase separation including solid and fluid model as special cases. Viscoelastic phase separation in dynamically asymmetric mixtures can be characterized by the order-parameter switching phenomena: The primary order parameter switches from the composition to the deformation tensor as in gel, and back to the composition again, reflecting viscoelastic relaxation between a characteristic deformation time of phase separation and the slowest rheological time of the system. This unusual behavior can be explained by the above general nature of a viscoelastic model. It is remarkable that order-parameter switching can occur even during an ordering process driven by a 'single' thermodynamic driving force. We argue that spongelike patterns observed in phase separation of many materials may be produced by a common physical mechanism of viscoelastic phase separation. For example, spongelike patterns formed during polymerization-induced phase separation can also be explained by our viscoelastic model.

Introduction

Phase-separation phenomena are generally observed in various kinds of condensed matter including metals, semiconductors, simple liquids, and complex fluids such as polymers, surfactants, colloids, emulsions, and biological materials. The study of the dynamic processes of phase separation is very important for both engineering applications and basic understanding of nonequilibrium dynamics of pattern formation [1]. On the basis of the concept of dynamic universality, types of phase separation are classified into various models by Hohenberg and Halperin [2]. For example, phase separation of solids is known as "solid model (model B)",

Y. Morishima · T. Norisuye · K. Tashiro (Eds.)
Molecular Interactions and Time-Space Organization in Macromolecular Systems
Proceedings of the OUMS '98, Osaka, Japan, 3-6 June, 1998

while phase separation of fluids as "fluid model (model H)" [2]. For the former the local concentration can be changed only by material diffusion, while for the latter by both diffusion and flow. A universal feature of the phase-separation kinetics and critical phenomena in each model has been established [1,2].

Here we focus our attention in phase separation in complex fluids that are characterized by the large internal degrees of freedom. In all conventional theories of critical phenomena and phase separation, the same dynamics for the two components of a binary mixture, which we call "dynamic symmetry" between the components has been implicitly assumed [1,2]. However, this assumption is not always valid especially in complex fluids. Recently we have shown that in mixtures having intrinsic "dynamic asymmetry" between its components (e.g. polymer solutions composed of long chainlike molecules and simple liquid molecules), a critical concentration-fluctuation mode is not necessarily the only slow mode of the system and, thus, we have to consider the interplay between critical dynamics and the slow dynamics of material itself [3]. We also found unusual phase separation, which we call "viscoelastic phase separation", in dynamically asymmetric mixtures such as polymer solutions [3] and polymer blends whose components have very different glass-transition temperature [4]. In addition to a solid and a fluid model, thus, we need the third general model for phase separation in condensed matter, which we call a "viscoelastic model" [5]. The relevance of our model to viscoelastic phase separation experimentally observed is confirmed by numerical simulation [6]. In this paper, we summarize the characteristic features of the "viscoelastic model" of phase separation. Then, we show that these special features of a viscoelastic model lead to a novel phenomenon of "order-parameter switching" during viscoelastic phase separation even if it is driven by a single thermodynamic force. Further, a possible common origin of spongelike structures observed in various materials will be discussed on the basis of our viscoelastic model. We argue that colloid phase separation and polymerization-induced phase separation can also be explained by the viscoelastic model.

Viscoelastic model of phase separation

Recently shear effects on complex fluids have attracted much attention because of its unusual nature known as "Reynolds effect": For example, shear flow that intuitively helps the mixing of components induces phase separation in polymer solutions [7]. There are many examples of such effects in complex fluids such as surfactant systems, block copolymers, and charged colloid systems. This is caused by the coupling between shear velocity fields and the elastic internal degrees of freedom of complex fluids. To explain this unique feature of polymer solution, there have been considerable theoretical efforts [8-12]. Doi and Onuki [12] derived generalized diffusion equations based on a two-fluid model, as an extension of the original idea of Brochard and de Gennes [8]. We have proposed [3] that viscoelastic phase separation of dynamically asymmetric mixtures can be described by their dynamic equations with some modification [5]. The differences between our problem and the problem of shear effects are as follows: In our case the velocity fields are self-induced by phase separation itself, while they are external shear fields in the problem of shear effects on phase separation. More importantly, bulk stress plays a crucial role in our case since it is directly coupled to diffusion, while it plays no role in the shear problem. The dynamic equations of the viscoelastic model are as follows [5]:

$$\frac{\partial \phi}{\partial t} = -\vec{\nabla}\cdot(\phi\vec{v}) + \vec{\nabla}\cdot\frac{\phi(1-\phi)^2}{\varsigma}[\vec{\nabla}\cdot\overleftrightarrow{\Pi} - \vec{\nabla}\cdot\overleftrightarrow{\sigma}^{(n)}]. \tag{1}$$

$$\vec{v}_p - \vec{v} = -\frac{(1-\phi)^2}{\varsigma}\vec{\nabla}\cdot[\overleftrightarrow{\Pi} - \overleftrightarrow{\sigma}^{(n)}]. \tag{2}$$

$$\rho_0\frac{\partial \vec{v}}{\partial t} = -\vec{\nabla}\cdot[\overleftrightarrow{\Pi} - \overleftrightarrow{\sigma}^{(n)}] - \vec{\nabla}p + \eta_s\vec{\nabla}^2\vec{v}. \tag{3}$$

Here $\vec{v}_s(\vec{r},t)$ and $\vec{v}_s(\vec{r},t)$ are the average velocities of polymer and solvent at point $\vec{r}$ and time t, and $v = \phi v_p + (1-\phi)v_s$ is the average velocity, ϕ is the composition of polymer, and $\overleftrightarrow{\Pi}$ is the osmotic stress tensor. Please note that $\nabla\cdot\Pi = -\phi\nabla\,\delta F/\delta\phi$. The ρ_0 is the average density, p is a part of the pressure, η_s is the solvent viscosity, and ς is the friction constant per unit volume. Here the free energy F is given by the following Flory-Huggins-de Gennes form [13]:

$$F = \int d\vec{r}[f(\phi) + \frac{K}{2}(\vec{\nabla}\phi)^2],$$

$$f(\phi) = k_B T[\frac{1}{N}\phi\ln\phi + (1-\phi)\ln(1-\phi) + \chi\phi(1-\phi)],$$

where N is the degree of polymerization of polymer, and χ is the interaction parameter between polymer and solvent. In the linear response regime $\overleftrightarrow{\sigma}^{(n)}$ is generally written as [5]

$$\sigma_{ij}^{(n)} = \int_{-\infty}^{t} dt'[G(t-t')\kappa_{ij}^{p}(t') + K(t-t')\vec{\nabla}\cdot\vec{v}_p(t')\delta_{ij}], \tag{4}$$

where κ_{ij}^{p} is the deformation velocity tensor given by $\kappa_{ij}^{p} = \partial_i v_{pj} + \partial_j v_{pi} - 2/3(\partial_k v_{pk})$ and $G(t)$ is a material function called the shear relaxation modulus. The last term coming from bulk relaxation modulus $K(t)$ is introduced by us recently [5,6] and plays a crucial role in viscoelastic phase separation. The reason why a system has a bulk modulus is that a transient gel is commonly formed in dynamically asymmetric mixtures just after a quench [5]. Note that $\eta = \int G(t)dt$, where η is the viscosity of polymer solution. Equation (2) means that the motion of polymer relative to the solvent is driven by not only the osmotic stress tensor $\overleftrightarrow{\Pi}$, but also the network stress $\overleftrightarrow{\sigma}^{(n)}$. This theory primarily deals with polymer solutions, but it should be applicable to any dynamically asymmetric mixtures by changing the expressions of $G(t)$, $K(t)$, and $f(\phi)$ to the relevant ones. It should be noted that the rheological relaxation functions $G(t)$ and $K(t)$ are functions of local composition $\phi(\vec{r})$ and, more importantly, we need some microscopic or phenomenological theories to estimate $G(t)$ and $K(t)$, which depend upon mechanical properties of the material.

Generality of Viscoelastic Model [5]

First we consider the general nature of the above basic equations describing viscoelastic phase separation. If we set $\vec{v}=0$ and $\vec{\sigma}^{(n)}=0$ in Eqs.(1)-(3), we obtain the basic equation of "solid model (model B [2])". If we set $\vec{v}=0$, $G(t)=\mu$, and $K(t)=\kappa$ (μ, κ : shear and bulk modulus), we obtain the basic equations of "elastic solid model" [14]. If we set only $G(t)=\mu$ and $K(t)=\kappa$, we obtain the basic equations of "elastic gel model". Note that the time integration of the velocity becomes the deformation u, and $\sigma_{ij}^{(n)} = \mu[\partial_j u_{pi} + \partial_i u_{pj} - 2/3(\partial_k \vec{u}_{pk})\delta_{ij}] + \kappa(\partial_k \vec{u}_{pk})\delta_{ij}$. If we assume slow deformation (κ_{ij}, $\nabla \cdot v_p \sim const.$), on the other hand, we obtain the basic equations of "fluid model (model H [2])". Thus, the viscoelastic model is a general model that can describe any types of phase separation in mixtures of isotropic condensed matter, as its special cases. The viscoelastic model in the classification of isotropic phase separation corresponds to viscoelastic matter in the classification of isotropic condensed matter. Viscoelastic matter includes any condensed matter ranged from solid to fluid. The key factor for the classification of material is the relation between the characteristic internal rheological time and the characteristic observation time. Corresponding to this, the key physical factor for the classification of isotropic phase separation is the relation between the characteristic time of phase separation and the characteristic rheological time of the slower phase. The above analogy is summarized schematically in Fig.1.

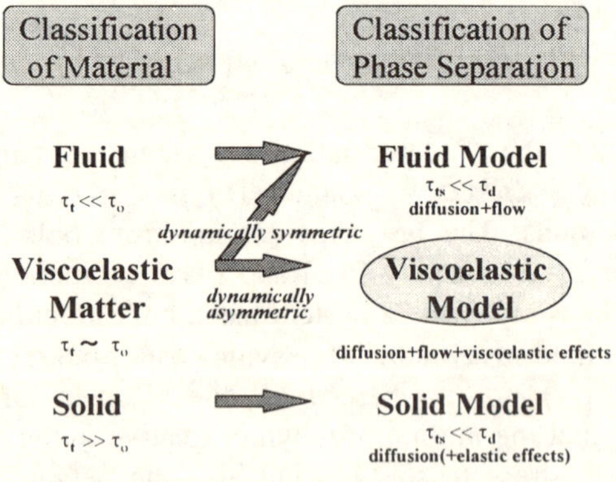

Fig. 1 A schematic figure explaining a new classification of phase separation of isotropic matter and its relation to the classification of material. τ_t is the characteristic rheological time, τ_o is the characteristic observation time, τ_{ts} is the characteristic rheological time of the slower phase, and τ_d is the characteristic time of domain deformation.

Order-parameter switching during viscoelastic phase separation

In the viscoelastic model, the phase-separation mode can be switched between "fluid mode" and "elastic gel mode". This switching is caused by the change in the coupling between stress fields and velocity fields, which is described by Eq. (4): Equation (4) tells us that these two ultimate cases, namely, (i) fluid model ($\kappa_{ij}^{p}, \nabla \cdot \vec{v}_p \sim const.$) and (ii) elastic gel model ($G(t), K(t) \sim const.$), correspond to $\tau_{ts} >> \tau_d$ and $\tau_{ts} << \tau_d$, respectively. Here τ_{ts} is the characteristic rheological time of the slower phase and τ_d is the characteristic time of domain deformation. For $\tau_{ts} >> \tau_d$ the primary order parameter is the composition in usual classical fluids, while for $\tau_{ts} \leq \tau_d$ it is the deformation tensor as in elastic gels. Thus, we can say that the order-parameter switching is a result of the competition between two time scales characterizing domain deformation τ_d and the rheological properties of domains τ_{ts}. This is a kind of *viscoelastic relaxation* in pattern evolution. We now consider how τ_{ts} and τ_d change with time during phase separation. In the initial stage, the velocity fields grow as $v \sim (k_B TK / 3\eta\xi)\Delta\phi^2$, where $\Delta\phi$ is the composition difference between the two phases, and ξ is the correlation length, or the interface thickness. Since $\Delta\phi$ approaches to $2\phi_e$ (ϕ_e :the equilibrium composition) with time, this expression of v reduces to the well-known relation $v \sim \gamma / \eta$ (γ : interface tension) in the late stage [note that $\gamma \sim k_B TK(2\phi_e)^2 / 3\xi$]. Thus, the characteristic deformation time τ_d changes with time as $\tau_d \sim R(t) / v(t) \sim R(t) / \Delta\phi^2$. In the initial stage, the domain size does not grow so much with time while $\Delta\phi$ rapidly increases with time; and, accordingly, τ_d decreases rapidly. On the other hand, τ_{ts} increases steeply with an increase in $\Delta\phi$, reflecting the increase in the polymer concentration in a polymer-rich domain. Thus, τ_{ts} becomes comparable to τ_d in the intermediate stage of phase separation. Once $\tau_d < \tau_{ts}$, the slower phase cannot follow the deformation speed and behaves as an elastic body: The elastic energy dominates the coarsening process in the intermediate stage. Next we consider the late stage. Since $\Delta\phi$ approaches to $2\phi_e$ and becomes almost constant in the late stage, $\tau_d (\sim R\eta / \gamma)$ increases with an increase in R while τ_{ts} becomes almost constant. Thus, τ_d becomes longer than τ_{ts} again. In short, $\tau_d >> \tau_{ts}$ in the initial stage, $\tau_{ts} \leq \tau_d$ in the intermediate stage, and $\tau_d >> \tau_{ts}$ in the late stage again. Accordingly, the order parameter switches from the composition to the deformation tensor, and then switches back to the composition again. This is likely the first case of "order-parameter switching" during an ordering process driven by a `single' thermodynamic driving force.

Because of this phenomenon, there is no self-similarity in pattern evolution of viscoelastic phase separation. In the elastic regime, further, the volume ratio between the two phases changes with time and thus there is no proportionality between interdomain distance and domain size. This is also related to the fact that there is no characteristic length scale for elastic deformation.

Special features of viscoelastic model

Here we point out that the viscoelastic model is not necessarily a universal model of critical phenomena in the practical sense and it intrinsically has nonuniversal nature: Theoretically, the only slow mode should be critical concentration fluctuation extremely near a critical point. In an experimentally accessible temperature region, however, viscoelastic effects can play a significant role even near the critical point for mixtures having strong dynamic asymmetry. Practically, thus, we probably need reconsideration on the concept of the universality class in critical phenomena [1,2] for dynamically asymmetric systems. The key point is that there is an additional slow internal mode besides a critical mode of the order parameter.

A common physical origin of spongelike pattern

Here we discuss the general nature of the spongelike morphology and its physical origin. It is known that gel undergoing volume shrinking phase transition forms the bubble-like structure [8,10]. The physical origin of the appearance of the honeycomb structure in polymer foams (e.g. polystyrene foam and urethane foam) is also likely similar to ours. All these processes have a common feature that the holes of a less viscoelastic phase (gas in foam, water in gels, a solvent in polymer solution, and so on) are nucleated to minimize the elastic energy associated with the formation of a heterogeneous structure in an elastic medium. Then, the more viscoelastic phase decreases its volume. This volume shrinking process is dominated by the transfer (diffusion or flow) of the less viscoelastic component in the more viscoelastic phase to satisfy the total mass conservation. Thus, the limiting process of pattern evolution is commonly the diffusion or flow process of the fluid component under stress. Among these examples, a polymer foam is unique in the point that the total volume is not conserved since it accompanies the liquid-gas phase transition. In addition to patterns in mixtures of condensed matters, we also point out the similarity of these patterns to the foamlike structure of the universe [11], where the gravitational attractive interaction likely plays a role similar to elasticity in producing the structure. A spongelike structure is probably the universal morphology for phase separation of a system in which only one component asymmetrically has elasticity stemming from either topological connectivity or attractive interaction. The common feature of these systems originates from the volume phase transition, or more strictly elastic phase separation of dynamically asymmetric mixtures including fluid (liquids and gasses) at least as a component. This differs from the elastic phase separation of dynamically symmetric solid mixtures (e.g. metal alloys) which does not accompany the volume change of each phase [4]: In this case, a softer phase always form a continuous phase to minimize the total elastic energy, in contrary to our case.

Further examples of viscoelastic phase separation

Hereafter we consider two examples of phase separation, in which we believe viscoelastic effects play a crucial role.

Phase separation in colloidal suspensions

Recently it was found [15-18] that the addition of non-absorbing polymers to a colloidal suspension can cause phase separation due to polymer-induced depletion attraction between colloidal particles. When colloids are close enough, there is an overlap of depletion zone from which polymers are sterically excluded. The resulting unbalanced osmotic force causes attractive interactions between colloidal particles. Depending upon the sizes of polymers and colloids and their compositions, the variety of phase separation behavior is reported [19-22], including (i) fluid-fluid phase separation, (ii) gellike phase separation, and (iii) phase-separation induced crystallization. For example, when colloidal suspensions are brought shallowly into an unstable region [case (i)], the early stage of phase separation cannot be described by the standard Cahn's linear theory and the transport coefficient apparently has a strong q dependence [20,21]. For a deep quench [case (ii)], on the other hand, the initial growth of the concentration fluctuations are followed by the formation of a transient gel, and the coarsening process apparently stops for a while. This transient gel state lasts for a long time, and then the gel eventually collapses under the gravity [19-21]. Three-dimensional microscopic observation reveals that big holes are slowly created in a transient gel during the above process [21]. Its final state can be well described by the thermodynamic phase diagram. The quite similar phenomena are also observed in phase separation of other types of colloidal suspensions [23] and also that of emulsions [24,25]. The study of these interesting phenomena just began recently, and the situation is still far from the complete understanding.

We argue that our viscoelastic model can explain the above unusual features of phase separation of colloidal suspensions and emulsions: These phenomena can be explained by "asymmetric stress division" between the components of a mixture due to their size difference. The interaction network of particles, which is responsible for the formation of a transient gel, can store elastic energy, while a fluid component cannot. The importance of the bulk stress stemming from interaction network can be explained by using a concept of "topological constraint for diffusion". We point out the universal features common to polymer solutions, colloidal suspensions, and emulsions, which are characterized by the concepts of "dynamic asymmetry" and "interaction network" [26].

Polymerization-induced phase separation

Polymerization-induced phase separation has recently attracted increasing attention from the both fundamental and industrial viewpoints. The origin of the phenomenon is very simple: Suppose there is a homogeneous mixture of two liquids. If we initiate the polymerization of one of the components (A) by light or heat, the contribution of the mixing entropy significantly decreases as the polymerization undergoes. Polymerization effects can be included in the theory as a temporal change in the degree of polymerization $N_A(t)$. According to the Flory-Huggins mean-field theory, the mixing free energy F can be given by

$$f(\phi) = k_B T[\frac{1}{N_A(t)}\phi \ln \phi + \frac{1}{N_B}(1-\phi)\ln(1-\phi) + \chi\phi(1-\phi)],$$

In some cases, a mixture finally becomes unstable and starts to phase separate. During phase separation, one of the components, A, is polymerized and thus this asymmetric growth of the molecular size leads to the strong dynamic asymmetry between the two components. In this polymerization-induced phase separation, there can be two types of new effects that may affect the pattern evolution; namely, viscoelastic and elastic effects. The former predicts that a more viscoelastic phase forms a spongelike structure for linear polymerization regime even when it is a minority phase. The latter, on the other hand, predicts that a softer phase forms a spongelike pattern for network-forming polymerization or gelation even when it is a minority phase. The viscoelastic effect is the kinetic effect, while the elastic effect is the static, or energetic effects. This difference is responsible for the above entirely different predictions. Pattern evolution of polymerization-induced phase separation is first strongly affected by the initial growth of dynamic asymmetry and the resulting viscoelastic phase separation, and then the structure is frozen by crosslinking reaction. Because of this kinetic reason, thus, we believe that the viscoelastic effects dominate the pattern evolution. Thus, we predict that a more viscoelastic phase, namely, a polymerized phase, forms a networklike or spongelike pattern during polymerization-induced phase separation. Actually, such morphology is commonly observed for polymerization-induced phase separation [27]. Thus, morphological evolution in polymerization-induced phase separation can also be well explained by our viscoelastic model.

Summary

We have demonstrated that viscoelastic phase separation can be generally observed in dynamically asymmetric mixtures such as polymer solutions, polymer blends, colloidal suspensions, and emulsions. It can lead to the formation of spongelike structure of a more viscoelastic phase even when it is a minority phase, contrary to the common belief of normal phase separation. From industrial viewpoints, this feature would be quite useful for making a percolated structure of a minority phase intentionally and controlling the mechanical properties. We argue that most of spongelike structures observed in various materials can be caused by viscoelastic phase separation. Spongelike patterns observed in polymerization-induced phase separation can also be explained by the above mechanism.

1. J. D. Gunton, M. San Miguel, and P. Sahni, in Phase Transition and Critical Phenomena, edited by C. Domb and J. H. Lebowitz (Academic, London, 1983) Vol.8.
2. P. C. Hohenberg and B. I. Halperin, Rev. Mod. Phys. **49**, 435 (1976).
3. H. Tanaka, Phys. Rev. Lett. **71**, 3158 (1993); J. Chem. Phys. **100**, 5253 (1994).
4. H. Tanaka, Phys. Rev. Lett. **76**, 787 (1996).
5. H. Tanaka, Phys. Rev. E **56**, 4451 (1997).
6. H. Tanaka and T. Araki, Phys. Rev. Lett. **78**, 4966 (1997).
7. see e.g. G. Ver Strate and W. Philippoff, Polym. Lett. **12**, 267 (1974); C. Rangel-Nafaile, A.B. Metzner, and K.F. Wissbrum, Macromolecules **17**, 1187 (1984); B. A. Wolf, Makromol. Chem. **189**, 163 (1988).
8. F. Brochard and P.G. de Gennes, Macromolecules **10**, 1157 (1977); F. Brochard, J. Phys. France **44**, 39 (1983).
9. E. Helfand and G.H. Fredrickson, Phys. Rev. Lett. **62**, 2648 (1989).
10. A. Onuki, Phys. Rev. Lett. **62**, 2427 (1989).
11. S.T. Milner, Phys. Rev. E **48**, 3674 (1993).
12. M. Doi and A. Onuki, J. Phys. II France **2**, 1631 (1992).
13. P.G. de Gennes, *Scaling Concepts in Polymer Physics* (Cornell Univ. Press, N.Y., 1979).
14. see e.g. A. Onuki and H. Nishimori, Phys. Rev. B **43**, 13649 (1991).
15. W. B. Russel, D. A. Saville, and W. R. Schowalter, *Colloidal Suspensions* (Cambridge Univ. Press, Cambridge, 1989).
16. A. P. Gast, C. K. Hall, and W. B. Russel, J. Colloid & Interface Sci. **96**, 251 (1983).
17. H. N. W. Lekkerkerker, W. C. K. Poon, P. N. Pusey, and A. Stroobants, and P. B. Warren, Europhys. Lett. **20**, 559 (1993).
18. W. C. K. Poon, J. S. Selfe, M. B. Robertson, S. M. Ilett, A. D. Pirie, P. N. Pusey, J. Phys. II France **3**, 1075 (1993).
19. P. N. Pusey, A. D. Pirie, W. C. K. Poon, Physica A **201**, 322 (1993); W. C. K. Poon, A. D. Pirie, and P. N. Pusey, Faraday Discussion **101**, 322 (1993); W. C. K. Poon, A. D. Pirie, M. D. Haw, and P.N. Pusey, Physica A **235**, 216 (1997).
20. N. A. M. Verhaegh, J. S. van Duijneveldt, J. K. G. Duhont, and H. N. W. Lekkerkerker, Physica A **230**, 409 (1996).
21. N. A. M. Verhaegh, Ph.D. Thesis (1997); N. A. M. Verhaegh, D. Asnafhi, H. N. W. Lekkerkerker, M. Giglio, and L. Cipelletti, Physica A (in press).
22. R. M. L. Evans, W. C. K. Poon, and M. E. Cates, Europhys. Lett. **38**, 595 (1997).
23. M. Carpineti and M. Giglio, Phys. Rev. Lett. **68**, 3327 (1992); ibid. **70**, 3828 (1993).
24. J. Bibett, T. G. Mason, H. Gang, and D. A. Weitz, Phys. Rev. Lett. **69**, 981 (1992).
25. see, e.g., E. Dickinson and M. Golding, J. Colloid & Interface Sci. **191**, 166 (1997); Food Hydrocolloids **11**, 13 (1997).
26. H. Tanaka, submitted to Phys. Rev. E.
27. P.S. Drzaic, Liquid Crystal Dispersion (World Scientific,Singapore, 1995).

Dielectric Studies of Order-Disorder Transition in Solutions of Styrene-Isoprene Block Copolymers

Keiichiro Adachi

Department of Macromolecular Science, Graduate School of Science, Osaka University, Toyonaka, Osaka 560-0043, Japan
adachi@chem.sci.osaka-u.ac.jp

Abstract. We investigated the dielectric normal mode relaxation of the polyisoprene (PI) block chain of styrene-isoprene diblock copolymer (SI) and styrene-isoprene-styrene triblock copolymers (SIS) in semidilute and concentrated solutions. Solvent was a common good solvent toluene (TOL). The order-disorder transition (ODT) was reflected on the dielectric loss curves ε": In the ordered state, SI solutions exhibited a bimodal ε" curve but a single peak in the disordered state. In this paper we investigated the molecular mechanism of the bimodal relaxation spectra in the ordered phase using solutions of non-symmetrical SI, triblock copolymers SIS, and styrene-butadiene-isoprene triblock copolymers (SIB and SBI). In order to detect the ODT we also carried out the measurements of heat capacity on an SI/TOL system. The bimodal relaxation spectra in the ordered state was ascribed to the difference in mobility of the chain ends of the PI block. From the dielectric data for solutions of SIS, SIB, and SBI in TOL, we concluded that the high and low frequency peaks of the ε" curves can be ascribed to the motion of the end at the interface and that of the free end, respectively.

Introduction

For a chain possessing dipole moment aligned in the direction parallel to the contour of the chain, we can observe the normal mode relaxation due to fluctuation of the end-to-end vector. Such a chain is classified as a type-A polymer by Stockmayer.[1] Thus we can obtain quantitatively the average correlation time and the distribution of correlation times for the auto-correlation function of the end-to-end vector. In our previous studies, we demonstrated that the dielectric normal mode relaxation spectroscopy is a powerful tool for investigation of the fluctuation of the end-to-end vector of flexible polymer chains both in solution and in undiluted state.[2] Beside the dynamical information, we can determine the mean square end-to-end distance from the relaxation strength.[2,3]

For non entangled chains, the bead-spring model has been the most successful model for chain dynamics.[4,5] It was found that the normal mode relaxation time of various type-A polymers such as polyisoprene (PI) and poly(ε-caprolactone) (PCL) in dilute solutions conform well to the Rouse-Zimm theory. For entangled chains, de Gennes proposed that the auto-correlation function $\phi(t)$ of the end-to end vector is given by exactly the same form as the Rouse theory.[6,7] Therefore if the dynamics of entangled chains can be described by the tube model, the relaxation spectrum should be independent of concentration above the overlapping concentration C*. However, we found that the dielectric loss curves of semidilute solutions of polyisoprene broadens with increasing concentration. For bulk polyisoprene, we also found that the spectrum does not conform to the de Gennes model and that the dielectric normal mode relaxation time is proportional to $M^{3.7}$. This behavior is close to the well known empirical 3.4 law for the longest viscoelastic relaxation time but the exponent is higher than the power 3.0 predicted by the tube model.[2] Recently we demonstrated that the constraint release model proposed by Graessley explains the 3.4 power law using the blends of mono disperse PI and polybutadiene (PB).[8]

Although there are still many intriguing unresolved issues on the dynamics of homopolymers, we have directed our attention to the dielectric normal mode relaxation in multi-component systems composed of type-A and non-type-A polymers. We expect that the mobility of type-A chains in multi-component systems is more restricted than in the single phase. There is few theoretical works on the problem how the spatial perturbation affects the power spectrum on the dynamics of polymer chains.

Typical multi component system is block copolymers. When the components of a block chain are immiscible, the block polymer forms microphase structure. At low temperature, the block polymers are in a strong segregation state but turns into a weak segregation with increase of temperature.[12] Thus block chains confined in a narrow space of the microphase exhibit complicated dynamics which cannot be explained by the Rouse or the tube models. When one component is glassy, the block chain of the other component move in a narrow phase of the microphase with one end anchored at the interface of the micro domain.

Previously we reported that the normal mode relaxation spectra of bulk styrene-isoprene(SI) diblock copolymers are much broader than the homopolymer.[9-11] We also found that concentrated solutions of a symmetrical styrene-isoprene block copolymer (SI) in a common good solvent toluene (TOL) exhibited a bimodal dielectric loss curve below its order-disorder transition temperature ODT but a single peak above the ODT.[11] In this case the styrene phase is in the state of concentrated solution. Thus the junction is mobile at the interface. The SI is expected to have a lamella morphology in the ordered state. The transition was observed quite sharply both in the temperature and concentration dependence of the ε" curves.

In this paper, we investigated the time-space organization in semidilute solutions of SI block copolymers. There are two objectives in the present study. First is to clarify the dielectric behavior of semidilute solutions of asymmetric SI near the ODT. In order to detect the ODT, we also carried out the heat capacity measurement. The second objective is to clarify the mechanism of the bimodal relaxation spectrum in solutions of SI. One may consider that if the mobility of the two chain ends of the PI block are different, the relaxation spectrum becomes bimodal. In order to examine the mobility of the free end and the junction to the styrene block separately, we used SIS, styrene-butadiene-isoprene triblock copolymer SBI, and styrene-isoprene-butadiene triblock copolymer SIB in which the butadiene block does not possess the type-A dipole moment. In case of SIS, the both ends of the PI block are linked with the styrene blocks. Thus the dielectric response is due to the junctions. On the other hand, SBI and SIB exhibit dielectric relaxation due to the motion of the free chain end and the junction, respectively, since the I block locates at the free end and the junction, respectively.

Experimental Section

Samples

All polymer samples were prepared by anionic living polymerization using sec-butyllithium as the initiator. The samples have narrow distribution of molecular weight and their polydispersities M_w/M_n are less than 1.1. The code numbers of the samples indicate the weight average molecular weights of the blocks in kg/mol. For example SIB((53-22-49) is a block copolymer composed of S, I and B blocks with M_w = 53000, 22000, and 49000, respectively.

Method

Dielectric measurements were carried out with a transformer capacitance bridge (General Radio 1615A). Heat capacity was measured with an adiabatic type calorimeter. Details of the calorimeter will be described elsewhere.

Results and Discussion

1. Dielectric Behavior of SI with Symmetrical Composition

In our previous study, we reported that the dielectric normal mode relaxation spectra of the PI block chains in solutions of SI(42-42) in toluene were bimodal in the ordered state but unimodal above ODT.[11] In the case when the S and I contents are close to 50 %, the morphology is expected to be lamella.

Figure 1 shows the double logarithmic plot of the dielectric loss factor ε" vs frequency *f* for 20wt% solution of SI(76-67). Similar behavior was observed for 20 % solutions SI (92-80) as shown in Figure 2. It is seen that SI(76-67) also shows a bimodal ε" curve in the ordered state and a unimodal one in the disordered state. As is seen in those figures the bimodal spectra in solutions of symmetrical SI are independent of total molecular weight. The ODTs of the 20 wt% solutions of SI(42-42), SI(76-67), and SI(92,80) are 250, 280, and 270 K, respectively and depend on the total molecular weight and the composition. Generally the transition temperature of SI solutions increases with molecular weight. However, we see that the ODT of SI(76-67) is lower than that of SI(92-80). This indicates the ODT is very sensitive to the composition.

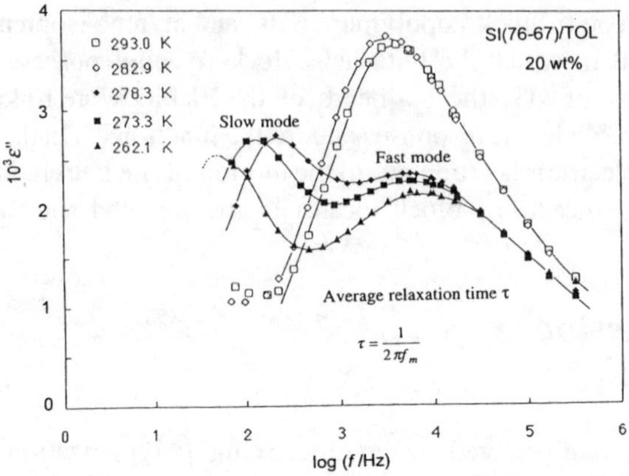

Fig. 1. The frequency dependence of 20 wt% solutions of SI(76-67) in toluene.

We also found that a slight change of concentration causes a large change in the ODT and the dielectric behavior. For instance, ODT of 22 wt % solution of SI(92-80) was observed at 301K and the shape of the loss curve just below the ODT was not bimodal but the high frequency mode was seen as a shoulder. With decreasing temperature the high frequency mode exhibited a peak below 290K.

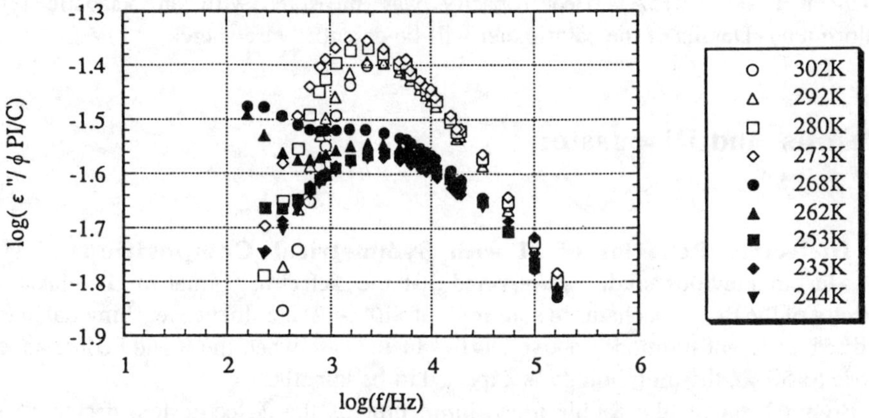

Fig. 2. log ε"/ϕPI/C vs log f plots for 20wt% solutions of SI(92-80). Here ϕPI and C are the PI content in the SI and polymer concentration, respectively.

Similar behavior was observed for 25% solution of SI(92-80). We expected the ODT is above 340K above which we could not carry out measurements due to the evaporation of toluene. At the concentration of 30 wt%, only the high frequency peak was observed in our experimental window.

2. Dielectric Behavior of SI Solutions having Cylindrical Morphology.

Figure 3 shows the temperature dependence of the normalized ε" curves of 20 wt% solutions of SI(27-78) which have probably a cylindrical morphology. In contrast to solutions with lamella morphology, this system does not exhibit clear bimodal spectrum. It is seen that the ε" curve broadens with decreasing temperature and rather suddenly around 255 K. We see that a small shoulder is seen in the high frequency side of the loss curve. If the broadening is due to the ODT we expect that the ODT of this system is ca. 255 K. When concentration was increased at 273, the loss curve also broadened with increasing concentration and at the same time it shifted to lower frequency side.

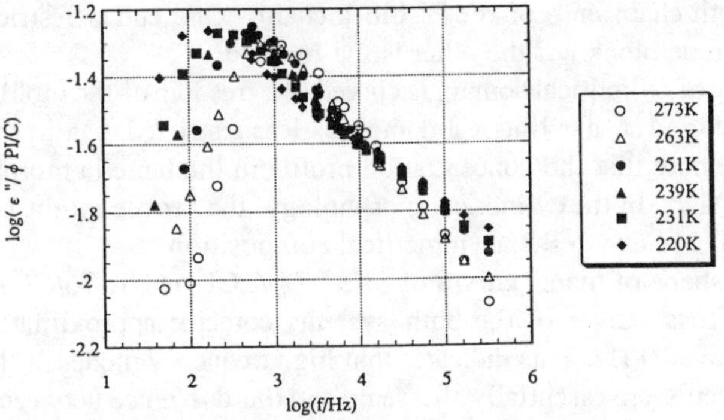

Fig. 3. log ε"/ϕPI/C vs log f plots for 20wt% solutions of SI(27-78).

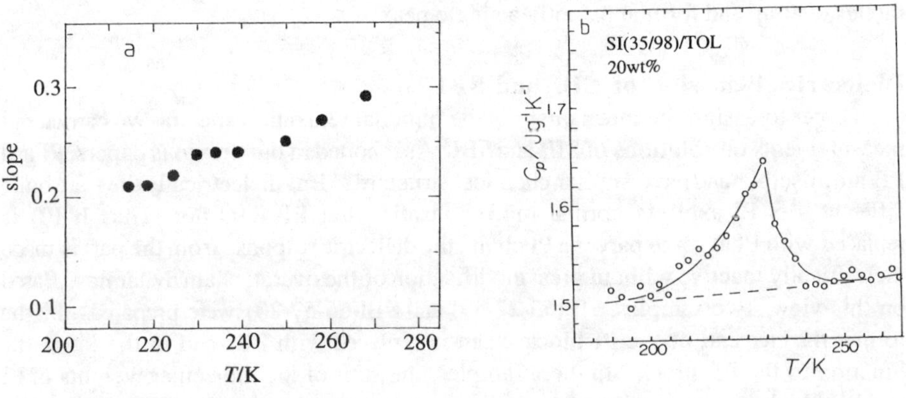

Fig. 4. (a) Slope of log ε" vs log f curve of 20 wt % solution of SI(36-98) and (b) temperature dependence of heat capacity of the same solution.

The ε" curves of 20 and 25% TOL solutions of SI(35-98) exhibited similar behavior to SI(27-78). The ε" curve broadened with decreasing temperature and no clear transition was observed. Figure 4a shows the temperature dependence of the slope of the log ε" vs log f curves in the high frequency side for the 20% solutions of SI(36-98).

It is seen that the slope increases stepwise around 230 K although the change is very small. Figure 4b shows the temperature dependence of heat capacity for 20wt% solution of SI(35-98) measured by an adiabatic calorimeter. The heat capacity curve indicates clearly a lambda shaped anomaly of the heat capacity and the anomaly is ascribed to the ODT. Thus it is clear that ODT occurs also in solutions of asymmetrical SI. Comparing those figures 4a and 4b, we may conclude that the small but sudden broadening of the ε" curve in SI(27-78)/TOL and SI(36-98)/TOL systems is due to ODT.

Here one may ask the reason of the difference in the dielectric relaxation spectra between the lamella and cylindrical morphologies: the former is bimodal while the latter is unimodal and has only a shoulder. The difference may be ascribed to the mobility of the both chain ends of the PI block chain. One end is restricted at the junction to the styrene block and the other is the free end.

Since the surface of cylindrical domain is curved, the freedom of the mobility of the PI block chain anchored at a cylindrical domain is less restricted than in the lamella case. Secondly we note that the concentration profile in the lamella morphology is sinusoidal near ODT.[12] In the cylindrical morphology, the profile might be different from the sinusoidal one due to the asymmetrical composition.

Comparing the shape of the ε" curves of SI(32-78)/TOL and SI(76/67) solutions, we found that the loss curves of the both systems coincide approximately in the frequency range above 10^4Hz. This indicates that high frequency modes in the lamella and cylindrical systems are essentially the same and the difference between the both systems is in the separation of the high and low frequency modes. Namely, the two modes in the lamella system are well resolved but in the cylindrical system the both modes overlap and form apparently a single peak.

Dielectric Behavior of SIB and SBI.

In order to assign the mechanism of the bimodal relaxation spectra, we carried out measurements on solutions of SIB and SBI. As reported in our previous papers, PI and PB are miscible and have similar chemical structure.[13] But dielectrically they are quite different, i.e. PI exhibits normal mode relaxation but PB does not. Thus if PB is replaced with PI at some part of a PI chain, the dielectric response from the part is made dielectrically inactive without great modification of the overall chain dynamics. Based on this view, two samples SIB(53-22-49) and SBI(66-57-20) were prepared. In the former the free end of the PI block chain is replaced with PB, and in the latter the junction to the PS block. In these samples, the sum of the molecular weights of PI and PB blocks are approximately similar to the molecular weight of the PS block. Thus we can compare the behavior of them with those of SI(76-67) solutions.

Figure 5 shows the ε" curves of 20% solution of SIB(53-22-49) in which the dielectric spectrum reflects the motion of the chains around the junction. Since the ODT of SI(76-67) solution is 280K at 20% concentration, we consider that 20% solution of SIB(53-22-49) is in the ordered state below 280 K. Now we compare the ε" curve of 20 wt% solution of SIB(53-22-49) with that of SI(76-67). As shown in Figure 5, the loss maximum frequency of SIB(53-22-49) locates around log f=2.5. On the other hand the low frequency peak of the SI(76-67) also locates in this frequency

Figure 5, the loss maximum frequency of SIB(53-22-49) locates around log f=2.5. On the other hand the low frequency peak of the SI(76-67) also locates in this frequency range. This indicates that the low frequency mode can be ascribed to the motion around the interface of the PI and PB domains.

In SBI(66-57-20), the PI block locates at the free end of the chain and hence its dielectric spectrum reflects the motion of the free end. For this solution we observed loss maximum frequency around 10 kHz at concentration of 20wt%. This corresponds to the high frequency peak of the 20wt% solution of SI(76-67). Thus the loss peak can be ascribed to the motion of the free end. To summarize the dielectric spectra of TOL solutions of the SIB and the SBI indicate clearly that the low and high frequency peaks observed for SI(76-67) solutions below ODT are ascribed to the motions of the junction and free end, respectively.

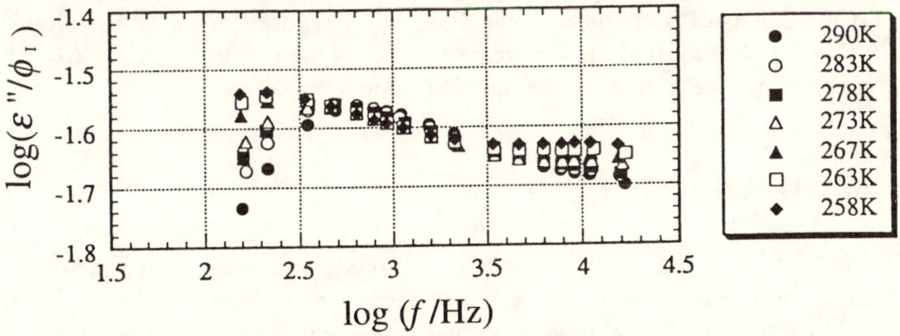

Fig. 5. Double logarithmic plot of ε''/ϕ_I against f for 20wt% solution of SIB(53-22-49) where ϕ_I represents the weight fraction of PI in the solution.

SIS Solutions

In the case of SIS triblock copolymers, both ends of the PI block are linked with the PS block chains. Thus the both ends are equivalent and their motions are expected to have similar relaxation time to the motion of the junction in SI diblock copolymers. In order to clarify the mechanism of the relaxation spectrum observed in SI diblock solutions, we also examined the dielectric behavior of SIS solutions.

Figure 6 shows the representative ε" curves of SIS(37-78-32) solutions with concentrations of 20 wt%. It is noted that the total molecular weight and the styrene content of this triblock copolymer is close to SI(76-67). Therefore we expect that SIS(37-78-32) solutions have a lamella morphology and that this system corresponds to solutions of SI(76-67). As shown in Figure 6, only a single peak is seen and shifts to lower frequency side with decreasing temperature. The fact that no drastic change in the temperature and concentration dependence of the ε" curve indicates that the system is in the ordered state. With further increase of the concentration, the peak shifted to low frequency side. The intensity and the shape are essentially the same as those of the 20 wt% solution. At 30 wt% solution, a new peak appeared in the high frequency range in low temperature range below 220 K. The peak is probably due to the segmental modes of the PS chains.

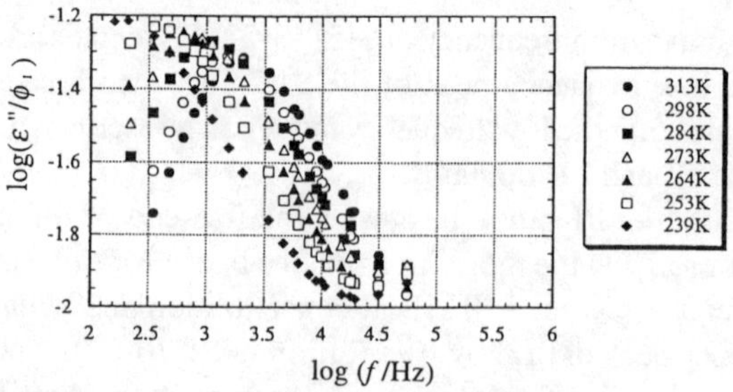

Fig. 6. Double logarithmic plot of ε''/ϕ_I against frequency f for 20 wt% solution of SIS(37-78-32).

At 20 wt% concentration the loss maximum is at log f= 2.9 at 273 K. This loss maximum frequency is close to the low frequency peak in the 20% solutions of SI(76-67) indication that the low frequency peak of the ε" curve of SI(76-67) solution corresponds to the motion of the junction point.

Conclusion

1. Dielectric normal mode process in semidilute solutions of symmetrical SI is bimodal in the ordered state but unimodal above the ODT.
2. Asymmetrical SI exhibits a normal mode relaxation spectrum with a small shoulder in the high frequency side of the loss curve. The shoulder disappears above the ODT.
3. The low frequency peak corresponds to the motion of the chain end linked to the styrene block and the high frequency peak to the motion of free end of the PI block.

References

1. W. H. Stoclmayer, Pure Appl. Chem. 15, 539 (1967).
2. K. Adachi and T. Kotaka, Progr. Polym. Sci. 18, 585 (1993).
3. K. Adachi, "Dielectric Spectroscopy of Polymeric Materials", Edt. J. P. Runt and J. J. Fitzgerald, Chapter 9, 261-282, American Chem. Soc.; Washington, D. C., 1997.
4. P. E. Rouse, J. Chem. Phys., 21, 1272 (1953).
5. B. H. Zimm, J. Chem. Phys., 24, 269 (1956).
6. P. G. de Gennes, J. Chem. Phys., 55, 572 (1972).
7. M. Doi and S. F. Edwards, J. Chem. Soc. Faraday Trans. 2, 74, 1789 (1978).
8 K. Adachi, T. Wada, T. Kawamoto, and T. Kotaka, Macromolecules, 28, 3588 (1995).
9. M.-L. Yao, H. Watanabe, K. Adachi, and T. Kotaka, Macromolecules, 24, 2955 (1991).
10 M.-L. Yao, H. Watanabe, K. Adachi, and T. Kotaka, Macromolecules, 24, 6175 (1991).
11 M.-L. Yao, H. Watanabe, K. Adachi, and T. Kotaka, Macromolecules, 25, 1699 (1992).
12 G. Bates and G. H. Fredrickson, Annu. Rev. Phys. Chem., 41, 525 (1990).
13 K. Adachi, I. Nishi, H. Doi, and T. Kotaka, Macromolecules, 24, 5843 (1991).

Long wavelength composition fluctuations in diblock copolymers

K. Chrissopoulou#, F. Rittig*, G. Fytas
Foundation for Research and Technology-Hellas
Institute of Electronic Structure and Laser
P.O. Box 1527, 711 10 Heraklion, Crete, Greece
E-mail: fytas@iesl.forth.gr

Abstract:
We examine the effect of composition polydispersity on the dynamic structure factor of diblock copolymers at low wavenumbers. Its bimodal relaxation shape in the case of a poly(styrene-b-isoprene) copolymer solution in the common good solvent toluene is preserved. The polydispersity is mainly manifested in the slow process associated with chain diffusion.

The most probable composition (order parameter) fluctuations ψ_q, in diblock copolymers AB are of mesoscopic length, O(R) (R being the polymer size) whereas long wavelength ψ_q are suppressed. For incompressible monodisperse AB, the static structure factor S(q) exactly vanishes at the thermodynamic limit q=0 and attains its maximum value for $x^* \equiv (qR)^2 \simeq 4$ (for a symmetric AB) [1]. It was soon recognized [2] that unavoidable polydispersity for real AB leads to a shift of $S(q^*)$ to lower q's and a fill-up of the correlation hole i.e. an excess $S_{exc}=S(q\rightarrow 0)\neq 0$. In the random phase approximation (RPA) [3], the dynamic structure factor S(q,t) = $<\psi_q(t)\ \psi_{-q}(0)>$ of monodisperse AB melts is a unimodal exponentially decaying function with a collective rate,

$$\Gamma_{coll} = \frac{xg(1,x)}{S(q)\tau_1} \tag{1}$$

where g(1,x) is the Debye form factor of the AB chain and τ_1 the longest chain relaxation time. In the presence of composition polydispersity κ_0, the total coherent $S_c(q,t)$ becomes a superposition of two exponential modes: S(q,t) and the incoherent $S_{exc}(q,t) = S_{exc}(q) \exp(-Dq^2t)$. In the absence of strong thermodynamic interactions, the additional diffusive contribution (polydispersity mode) is characterized by $D \simeq D_s$, the chain self diffusion coefficient and $S_{exc}(0) \propto \kappa_0$.

On the experimental side, the diblock copolymer dynamics, in contrast to their morphology, has been the subject of recent investigations using mainly dynamic light scattering. For either melts [4] or semidilute AB solutions [5], there is a consensus about the bimodal shape of $S_c(q,t)$ and the nature of the two thermal decay rates. Over the investigated reduced space $0<x<1$, where $x \equiv (qR)^2$, the

---

Also at the University of Patras, Physics department

* Perm. add.: Fakultät für Physik und Geowissenschaften, Universität Leipzig, D-04103 Leipzig, Germany.

dynamic structure factor of AB copolymers exhibits a pure structural relaxation (q-independent) and a slower diffusive (q^2-dependent) rate:

$$S(q) \propto q^2 R^2, \ \Gamma_{coll} \simeq \tau_1^{-1} \quad \text{and} \quad S_{exc} \propto \kappa_0, \ \Gamma_2 \cong D_s q^2 \qquad (2)$$

Note that access to the different contributions to the static structure factor is possible only in a dynamic scattering experiment. The feasibility of this dynamic resolution of the composition fluctuations with $q<R^{-1}$ has two relevant implications: (i) additivity of the total static structure factor probably as a result of uncoupled chain relaxation and diffusion and (ii) identification of the two normal modes of $S_c(q,t)$. Hence, it appears that for polydispersitiy typical of an anionic polymerization ($\kappa_0<0.01$) the low q fraction of the S(q,t) of ideally monodisperse AB can be unambiguously studied. However, the validity of these two findings for higher q's approaching q^* is not clear yet [6]. Its verification would require higher molecular weights (and hence R) with larger inherent polydispersity. Note that the necessity of a broad time range for a thorough $S_c(q,t)$ study makes the use of the photon correlation spectroscopy with light scattering q's (between 3×10^{-3}-4×10^{-2} nm^{-1}) unique.

In the present short report, we examine the polydispersity effect on the validity of the low q relaxation behavior of $S_c(q,t)$ at low q's (eq. 2). We have used a symmetric poly(styrene-b-isoprene) (SI) block copolymer (kindly provided by Prof. K. Adachi) with molecular weight M_w=172k and M_w/M_n=1.03. Note that this molecular weight is sufficiently large to resolve the chain relaxation in the disordered state (the order disorder transition (ODT) occurs at 20^0C for SI concentrations in toluene around 20wt%). Since the main cause of polydispersity in AB diblocks is incomplete removal of homopolymer, we have considered an SI/S mixture consisting of the same parent SI and 10wt% polystyrene with M_w=200k.

Figure 1, shows the net field correlation function, $C(q,t) = [(G(q,t) - 1) / f^*]^{1/2}$, (G is light scattering intensity autocorrelation function and $f^*<1$ is an instrumental factor), at q=0.034 nm^{-1} for the parent SI and the SI/S mixture (w_{ps}=10%) in toluene at total polymer concentration c~15wt% and 20°C. The analysis of the experimental C(q,t) usually proceeds via inverse Laplace transformation,

$$C(q,t) = \int L(\ln\tau) \exp(-t/\tau)\, d\ln\tau \qquad (3)$$

which allows the determination of the relaxation rate Γ_i and its intensity $I_i(=\alpha_i I)$ of the partitioning processes; I is the total light scattering intensity (normalized to that of neat toluene at the same scattering angle) and α_i is the amplitude of the ith-process in the distribution relaxation function L(lnτ).

The fastest diffusive process (Fig.1) is expected and found to be associated to the relaxation of the total polymer concentration fluctuations via cooperative diffusion. The other two processes are due to the composition fluctuations i.e. differences in the local concentrations of the two components (styrene, isoprene) and define the $S_c(q,t)$. For comparison reason the distribution functions $L(\ln\tau)$ are shown in the normalized form $IL(\ln\tau)$ in the inset of Fig. 1. At a first glance, it becomes clear that the slow polydispersity process dominates $C(q,t)$ ($\propto S_c(q,t)$) of the SI/S mixture and is responsible for the enhanced amplitude of the latter whereas the structural relaxation (intermediate process) remains virtually the same in both SI and SI/S solutions. Further the relaxation time (at the peak of $L(\ln(\tau))$ of the slow process clearly becomes slower in the SI/S mixture.

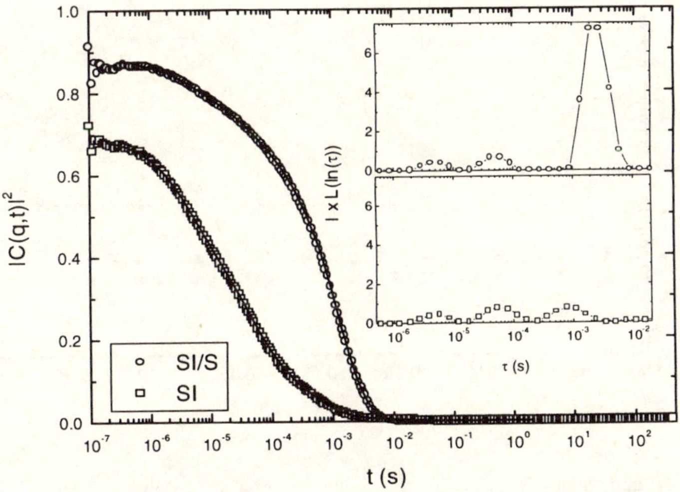

Fig.1: Experimental net correlation functions $C(q,t)$ at a scattering angle of 150^0 ($q=0.034nm^{-1}$) for SI (□) and SI/S (○) with total polymer concentration 15wt% in toluene at 20^0C. The lines are the fits of the ILT whereas the corresponding distribution of relaxation times $L(\ln(\tau))$ multiplied with the total scattered intensity $I(q)$ are plotted in the insets.

It is known that block copolymer/homopolymer blends exhibit both macrophase (at $q=0$) and microphase (at finite $q=q^*$) separation depending on the homopolymer content and the molecular weight of the two components. According to the theoretical phase diagram of AB/B blends [7] with $N_A \sim N_B$, the present SI/S mixture is predicted and found to exhibit microphase separation. However, the long wavelength ($q \to 0$) composition fluctuations become enhanced in the SI/S system due to the large composition polydispersity κ_0 and the unfavorable interactions ($\chi>0$) according to the expression [8]

$$S_{exc}(0) = \kappa_0 \phi N / \left[1 - 2\kappa_0 \chi N \phi^{1.59}\right] \tag{4}$$

derived for the excess low-q intensity of compositionally polydisperse AB's. In fact, the intensity of the slow mode (inset of Fig.1) is almost an order of magnitude higher in the SI/S system than in the parent diblock copolymer solution. The presence of homopolymer modifies $S_c(q,t)$ at long times, but it does not alter its bimodal shape (eq. 2) for $x<x^*$.

Figure 2 shows the characteristic time scales of the two relaxations (Fig.1) of the composition fluctuations at different reduced wavevectors in SI and SI/S systems.

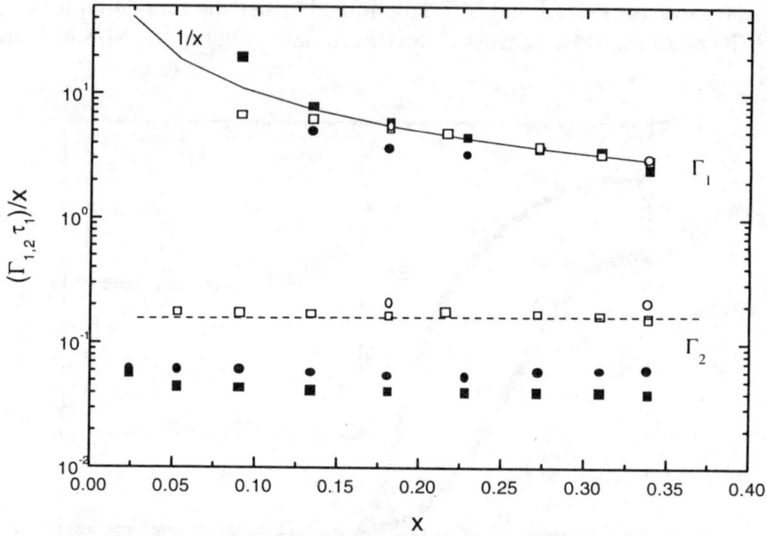

Figure 2: Reduced rates $(\Gamma_{1,2}\tau)/x$ for the collective chain (Γ_1) and diffusion (Γ_2) relaxation rate in SI (open symbols) and SI/S (solid symbols) solutions at two ((○,●) 15wt%, (□,■) 19wt%) total polymer concentrations in toluene at 20^0C. Solid line denotes the prediction of eq.1 whereas the dashed line is to guide the eye.

In order to compare the dynamics under isofriction conditions for different reduced length scales x, Fig.2 displays the rates in the form $\Gamma_{coll}\tau_l/x$ and $\Gamma_2\tau_l/x$. These reduced dynamic quantities should be insensitive to temperature and concentration variations far from the ordering transition as mobility effects are minimized. The data used in Fig.2 refer to two different copolymer concentrations (15 and 19wt%) at 20°C. The RPA prediction ($\propto g(1,x)/x$ for monodisperse diblocks in Eq.1) plotted as solid line is in agreement with the experimental Γ_l of the parent SI and the SI/S solutions. This finding indicates that the diblock copolymer dynamics for $x<x^*$ can still be resolved even in the presence of reasonably high polydispersity virtually reflecting the characteristic behavior of monodisperse AB's.

On the other hand, polydispersity affects the incoherent scattering contribution at low q's. Besides the enhancement of this intensity (Fig.1, eq.4) , the corresponding rate Γ_2 slows down in the SI/S solution as shown in Fig.2. While for SI, $\Gamma_2\tau_l/x\sim0.2$ (for two SI concentrations), for the SI/S solution $\Gamma_2\tau_l/x\sim$ 0.06 and 0.04 respectively for 15 and 19wt% total polymer concentration. Like the polydispersity effect on the excess low q scattering (eqs 2 and 4), the diffusion

$D = \Gamma_2/q^2$ associated with the thermal decay of the composition fluctuations, is also affected by the unfavorable interactions:

$$D = D_s[1 - 2\kappa_0 \chi N \phi^{1.59}] \quad (5)$$

Based on eqs 4 and 5, κ_0 in the SI/S is about three times larger than in the parent SI solution.

Concluding, the relaxation scheme for the bimodal dynamic structure factor of diblock copolymers for $q<q^*$ is valid in the presence of reasonably high polydispersity. The effect of the latter is exclusively manifested in the slow relaxation associated with chain diffusion.

Acknowledgment: F. Rittig thanks DAAD for the financial support.

References

[1] Leibler, L. Macromolecules 1980, 13, 1602.

[2] Burger, C.; Ruland, W.; Semenov, A. N. Macromolecules 1990, 23, 3339

[3] Akcasu, A. Z.; Benmouna, M.; Benoit, H. Polymer 1986,27, 1935. Erukhimovich, I. Ya.; Semenov, A. N. Zh. Eksp. Teor. Fiz. 1986, 63, 259 [Sov. Phys. JETP 1986,28,149]

[4] Anastasiadis, S. H.; Fytas, G.; Vogt, S.; Fischer, E. W. Phys. Rev. Lett. 1993, 70, 2415. Vogt, S.; Anastasiadis, S. H.; Fytas, G.; Fischer, E. W. Macromolecules 1994, 27, 4335. Stepanek, P.; Lodge, T. P.; Bates, F. S. Polym. Prepr. (Am. Chem. Soc., Div. Polym. Chem.) 1994, 35 (1), 616.

[5] Liu, Z.; Pan, C.; Lodge, T. P.; Stepanek, P. Macromolecules 1995, 28, 3221. Pan, C.; Mauer, W.; Liu, Z.; Lodge, T. P.; Stepanec, P.; von Meerwall, E. D.; Watanabe, H. Macromolecules 1995, 28, 1643. Jian, T.; Anastasiadis, S. H.; Semenov, A. N.; Fytas, G.; Adachi, K.; Kotaka, T. Macromolecules 1994, 27, 4762. Borsali, R.; Fischer, E. W.; Benmouna, M. Phys. Pev. A 1991, 43, 5732. Duval, M.; Haida, H.; Lingelser, J. P.; Gallot, Y. Macromolecules 1991, 24, 6867. Tsunashima, Y.; Kawamata, Y. Macromolecules 1994, 27, 1799.

[6]Boudenne, N.; Anastasiadis, S. H.; Fytas, G.; Xenidou, M.; Hadjichristidis, N.; Semenov, A. N.; Fleischer, G. Phys. Pev. Lett. 1996, 77, 506. Semenov, A. N.; Anastasiadis, S. H.; Boudenne, N.; Fytas, G.; Xenidou, M.; Hadjichristidis, N ; Macromolecules 1997, 30,.6280. Chrissopoulou, K.; Rittig, F.; Anastasiadis, S. H.; Fytas, G.; Fleischer, G.; Hadjichristidis, N. Manuscript in preparation.

[7]Matsen, M.W. Macromolecules, 1995, 28, 5765.

[8] Jian, T.; Anastasiadis, S. H.; Semenov, A. N.; Fytas, G.; Fleischer, G.; Vilesov, A. D. Macromolecules, 1995, 28, 2439. Molina, L. A.; Freire, J. J. J. Chem. Phys. 1998, 109, 2904.

Structure and Ordering Processes of Diblock Copolymer Associates in Solution

Takuhei Nose

Department of Polymer Chemistry, Tokyo Institute of Technology
O-okayama, Meguro-ku, Tokyo 152-8552, JAPAN
tnose@polymer.titech.ac.jp

Abstract. Experimental observations for micellar solutions of diblock copolymers are presented to show non-monotonous changes of association number during micellization processes, and reveal that some structural changes can be induced by completely kinetic factors. These findings are basically explained by the Aniansson-Wall (A-W) mechanism, which assumes only insertion/expulsion of unimers into/from micelles. It is demonstrated that phenomenological thermodynamic equations based on the A-W mechanism possibly reproduce a minimum in change of the association number with time observed in the relaxation process induced by the temperature jump. Furthermore, for more quantitative descriptions of kinetic behavior of the micellar solution, the time evolution of association-number distribution during micelle formation and relaxation is numerically calculated by evaluating the micelle free energy and the rate of expulsion of a unimer from micelles. The results clearly show the characteristic feature of the A-W mechanism, where the micellar relaxation takes place stepwise with two steps.

Introduction

Changes of temperature and/or concentration of micellar solutions sometimes induce changes of micelle shapes, for example, from spherical to cylindrical ones. Such changes have commonly been observed in conventional low molecular weight surfactant micelles. Block copolymer micelles, however, often show more complex association behaviors than the conventional surfactant micelles. For instance, Price et al.[1] and Stejskal et al.[2-4] have found that some block copolymer micelles formed on heating in solution are different from those on cooling. Eisenberg et al.[5,6] have revealed morphological complexity of diblock copolymer aggregates in solution depending on the composition of mixed solvent and the preparation procedures of solution. These experimental results suggest that observed micelles in polymeric systems are not always equilibrium ones, but appear during the micellization and/or relaxation processes of micelles. Actually, our recent studies on micellar solutions of a diblock copolymer have more directly

Y. Morishima · T. Norisuye · K. Tashiro (Eds.)
Molecular Interactions and Time-Space Organization in Macromolecular Systems
Proceedings of the OUMS '98, Osaka, Japan, 3-6 June, 1998

demonstrated that micelle size and shape have the process dependency and that some micelle structure is just formed kinetically.[7,8] Therefore, the kinetic aspect is indispensable for understanding and manipulating size and shape of obtainable micelles as well as other basic factors such as block lengths, copolymer architecture, and solvent selectivity.

Kinetics of micelle formation in diblock copolymer solution may basically be explained by Aniansson-Wall (A-W) theory, [9-13] where only insertion/expulsion of unimers into/from micelles and no fusion (or coalescence)/fission of micelles are assumed to be elemental processes. The theory leads to the presence of two distinguished processes, the fast and the slow processes. The fast process is accompanied with change in the number of unimers under a constant number of micelles, and the following slow process involves change of the total number of micelles with almost no change of the unimer number. The change of micelle number implies birth or extinction of micelles, which takes a long time under the condition of no excess free unimers. Such kinetics for micelle formation and relaxation by the unimer insertion and/or expulsion is called A-W mechanism.

In this presentation, first, we present experimental observations[7,8,12,13] showing non-monotonical changes of association number during micellization and the relaxation processes of micellar solutions, and that the structural changes can be induced by completely kinetic factors. Then, we will show that the non-monotonical changes observed can essentially be understood by the A-W mechanism. We apply phenomenological thermodynamic equations based on the A-W mechanism to show that the relaxation process by a temperature jump possibly exhibits a minimum in change of the association number with time. Furthermore, on the basis of the A-W theory, the time evolution of association-number distribution in micelle formation of diblock copolymer in solution is numerically calculated to understand kinetic behavior of the micellar solution.[14]

Basic Processes of Micellization

There are two basic micellization processes; (1) micellization from unimer state and (2) micellar relaxation induced by a temperature jump from one micellar state to another. To see kinetics of these basic processes, we have investigated micellization and relaxation kinetics of micelle-forming poly(α-methylstyrene)-*block*-poly(vinylphenethylalcohol)(PαMS-*b*-PVPA) in selective solvent, benzyl alcohol (BA), by means of time-resolved light scattering.[12,13] The copolymers, where PαMS block is much larger than PVPA block, form crew-cut type micelles with PαMS cores in this selective solvent.

Micellization Time evolutions of apparent molecular weight M_{wapp} and radius of gyration R_{gapp} reveal that the micellization from unimer state proceeds stepwise by fast and slow processes. The observed micelle-forming processes are illustrated in Figure 1a and are basically consistent with predictions of the A-W theory. The time constants for the fast and slow processes, which are estimated from the time evolution of M_{wapp}, exhibit characteristic concentration dependencies, respectively.

That is, the time constant τ_1 of the fast process decreases with increasing copolymer concentration, while the time constant τ_2 of the slow process is almost independent of the concentration, as shown in Figure 1b.

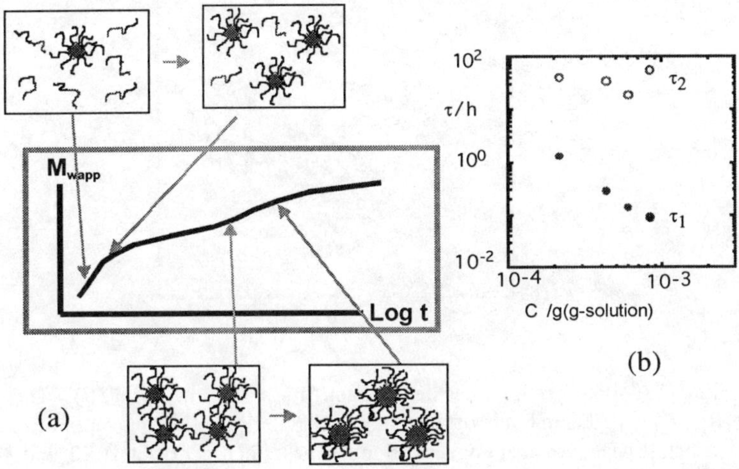

Fig. 1. Schematic illustrations of time evolution of the assosiation number and the association mechanism in the micellization from unimer state, and concentration dependencies of the time constants[7] for the fast process(τ_1) and slow process(τ_2).

Micellar relaxation The following two types of T-jumps were performed: (1) T-jump I from shallow T-depth to deep T-depth with the T-depth ΔT being defined by $\Delta T =$ cmt - T (cmt: the critical micelle temperature), and (2) T-jump II from deep ΔT_0 to shallow ΔT_1 with $\Delta T_0 > \Delta T_1$. In T-jump I, the micellar relaxation simply proceeds with monotonous increase of micelle size, while in T-jump II, deep-to-shallow quench, micelle size and micelle fraction do not monotonously change from the starting state to the final state, but the micelles first decompose with dominantly decreasing the micelle size, and then the micelle size and number recover to approach those at the final state. Time evolutions of M_{wapp} and R_{gapp} in T-jump II are shown as a function of concentration in Figure 2. The micelle decomposition at the first stage is enhanced with decreasing concentration.

Process-dependent Micelle Structure

Recently, we have found for polystyrene-*block*-poly(dimethylsiloxane) in the mixed selective solvent, *n*-octane / methylcyclohexane, that the micelle can take either a hollow spherical structure or a hollow cylindrical structure, depending on the condition of micelle formation.[7,8] This micellar solution has three different regions appearing *on heating* after quenching from unimer state, as illustrated in Figure 3a. Region I is the so-called anomalous-micellization region. In Region II, stable vesicles with the association number of about 10^3 are formed. Region III at *lower* concentrations is the region of the large hollow cylindrical micelles, which does not appear on cooling from unimer state.

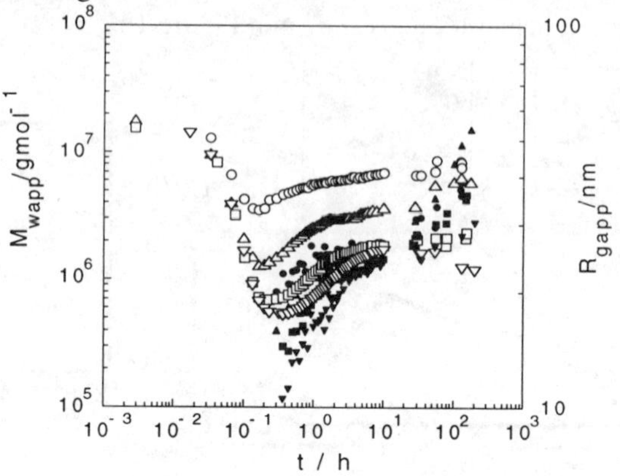

Fig. 2. Changes of apparent molecular weight(M_{wapp}) (hollow symbols) and radius of gyration(R_{gapp}) (filled symbols) of micelles under the micellar relaxation from deep-to-shallow quench.[8] The polymer concentration decreases in order of O, Δ, □, and ▼.

A time-resolved light scattering study has revealed that the large cylindrical micelles are not thermodynamically stable ones, but temporally formed in the course of micelle formation. This is clearly demonstrated by behaviors of micellization from unimer state at different temperatures and concentrations in Region II, which are illustrated in Figures 3b and 3c, along with estimated shape change during micellization (Figure 3d). The micellization does not monotonousy proceed to reach the destination, but passes through the formation of cylindrical micelles having association numbers larger than those of both initial and equilibrium micelles. At lower temperatures and lower concentrations, the cylindrical micelles with larger size and longer time stability are easily formed. The formation of the large cylindrical micelles is probably the result of a rapid growth of micelles by elongating into the cylindrical shape, which may come from bilayer structure of the micelles.

Theoretical Understanding and Description

As is shown in Figure 2, in the micellar relaxation from deep-to-shallow quench, the micelles first decompose instead of monotonous change to the final ones. To describe and understand this behavior on the bases of the A-W mechanism, we adopt the following thermodynamical equations for the association number, P, and the number of micelles, M, in solution.[8]

$$\frac{dP}{dt} = -L_p\left(\frac{\partial A}{\partial P}\right)_M \tag{1}$$

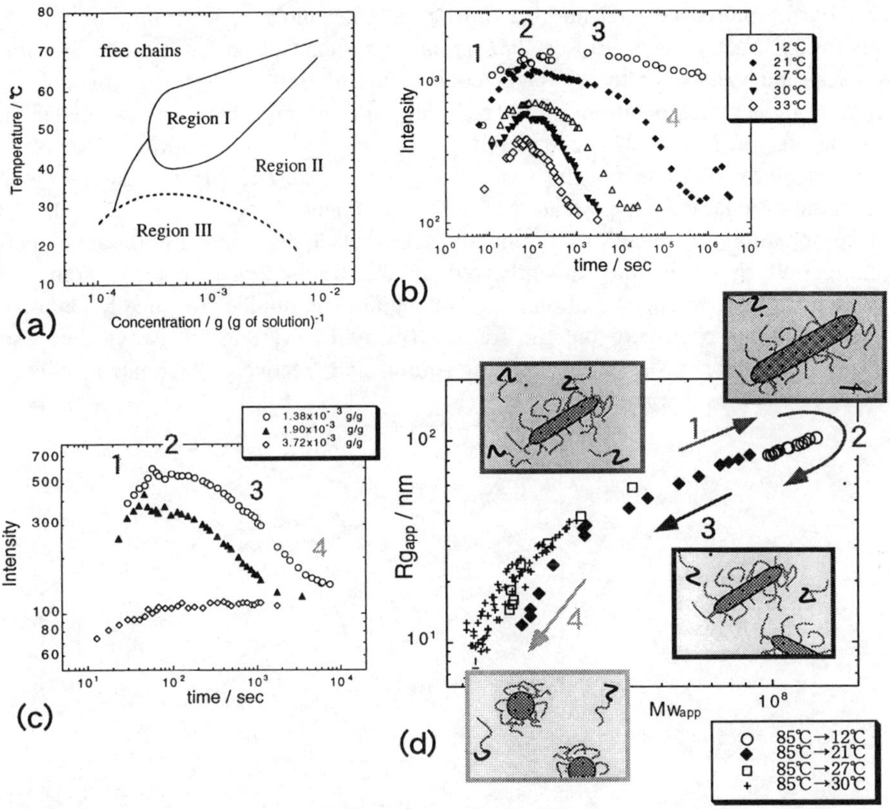

Fig. 3. Schematic presentations of micellization behavior of of PS-block-PDMS in C8 / MCH. (a) temperature-concentration diagram obtained on heating; (b) temperature and (c) concnetration dependencies of M_{wapp} vs. time; (d) shape change.

$$\frac{dM}{dt} = -L_M\left(\frac{\partial A}{\partial M}\right)_P \tag{2}$$

One can rewrite these equations in terms of the affinities A_P and A_M for P-change and M-change as

$$\frac{dP}{dt} = -\overline{L}_P(A_P - A_u)M \tag{3}$$

$$\frac{dM}{dt} = -\overline{L}_M(A_M - A_u)P \tag{4}$$

with the affinity A_u for unimers given as $A_u = \ln \phi_u$, where ϕ_m and ϕ_u are the volume fractions of micelles and unimers in solution, respectively, and $\overline{L}_P$ and $\overline{L}_M$ are reduced Onsager coefficients. At equilibrium, $A_P = A_M = A_u$ The affinities A_P and A_M can generally be evaluated as functions of P and M with their equilibrium values, without any particular model. Note that a set of P and M determines ϕ_m(or

ϕ_u). Furthermore, the total number N_t of copolymer molecules in solution, which is given by $N_t\phi_u$ / (ϕ_m + ϕ_u) + PM is always constant at any time. Providing appropriate values for the Onsager coefficients, the critical micelle concentration (cmc) and the association number P_e at equilibrium, one can calculate time evolutions of P and M, accordingly, those of any other quantities. Results of numerical calculations for the case of T-jump II are shown in Figure 4, which reproduce characteristic features of experimental observations well. The decomposition of micelles dominantly takes place by reducing the association number P in order to efficiently reduce the free energy of the system with increasing unimer concentration, since change of the number of micelles is a slow process. It is confirmed that the free energy of the system is always decreasing over the entire process of decomposition and recovery of micelles as the thermodynamics demands.

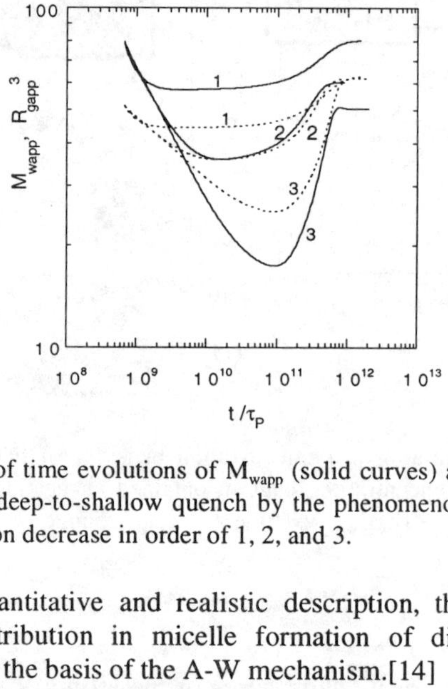

Fig. 4. Calculated curves of time evolutions of M_{wapp} (solid curves) and R_{gapp}(dotted curves) for micellar relaxation of deep-to-shallow quench by the phenomenological thermodynamic theory.[8] The concentration decrease in order of 1, 2, and 3.

To have a more quantitative and realistic description, the time evolution of association-number distribution in micelle formation of diblock copolymer in solution is calculated on the basis of the A-W mechanism.[14]
In the A-W theory,[9-11] we suppose the following process as the elemental *rate-determining* process for the change of association. That is, suppose the reaction that an associate A_p is formed from an associate A_{p-1} and a unimer A_1, and is decomposed into $A_{p-1}$and A_1, shown below.

$$A_{p-1} + A_1 \underset{b_p}{\overset{a_{p-1}}{\rightleftarrows}} A_p \qquad (p = 2, \cdots)$$

Define the flux J_p for net change in the number N_p of micelles A_p due to the association number change by the above reaction (from p-1 to p or p to p-1). Then, kinetic equations for the association reaction are written in terms of the flux J_p as

$$\frac{dN_1}{dt} = -2J_2 - \sum_{p>2} J_p \qquad (5)$$

$$\frac{dN_p}{dt} = J_p - J_{p+1} \quad p \geq 2 \tag{6}$$

The flux J_p is given by

$$J_p = a_{p-1} N_1 N_{p-1} - b_p N_p \tag{7}$$

with the rate constants a_{p-1} and b_p. Noting that $J_p = 0$ at equilibrium, one can rewrite equation (7) in terms of the equilibrium p-distribution $\overline{N}_P$ to

$$J_p = b_p \overline{N}_p \left[\frac{N_1}{\overline{N}_1} \cdot \frac{N_{p-1}}{\overline{N}_{p-1}} - \frac{N_p}{\overline{N}_p} \right] \tag{8}$$

A set of equations (5) and (6) is the fundamental equations determining the time evolution of the distribution N_p. To explicitly solve these equations, we have to know the free energy $f(p)$ of a micelle in solution, which can determine the equilibrium distribution $\overline{N}_P$, and the rate constant b_p of the elemental process of a chain expulsion from a micelle as a function of the association number p of the micelle.

In the present treatment, the scaling theory of Leibler et al.[15] is adopted to have an explicit expression of the free energy $f(p)$. On the other hand, the rate constant b_p is evaluated as a function of p by Halperin's treatment[16-18] based on the Kramers' rate theory,[19] where is supposed an expulsion process with the free energy vs. reaction coordinate that is illustrated in Figure 5. Numerical calculations are carried out for both cases of the micellization and the micellar relaxation. Examples of the results are shown in Figures 6 and 7.

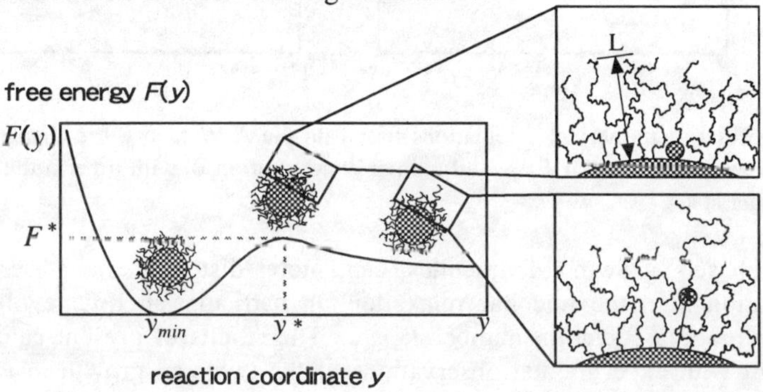

Fig. 5. Schematic illustrations for the chain-expulsion process and its free-energy profile as a function of reaction coordinate.

In the micellar relaxation of a shallow-to-deep T-jump (Figure 6), the micelle size changes stepwise with two steps, clearly showing the characteristic feature of the A-W mechanism, where there exist two processes, the fast process undergoing by consuming free unimers and the slow process accompanied with almost no change of unimer concentration. On the other hand, in the micellization from unimer state (Figure 7), the very fast process is observed, where the free chains get together quickly to form temporal micelles, and is followed by an ordinary micellar relaxation. As the copolymer concentration becomes higher, the very fast process goes on faster, while the rate of the fast process does not change. As the result, at lower concentrations close to the cmc, the very fast process merges with the fast process. The slow process has almost no concentration dependence in its rate.

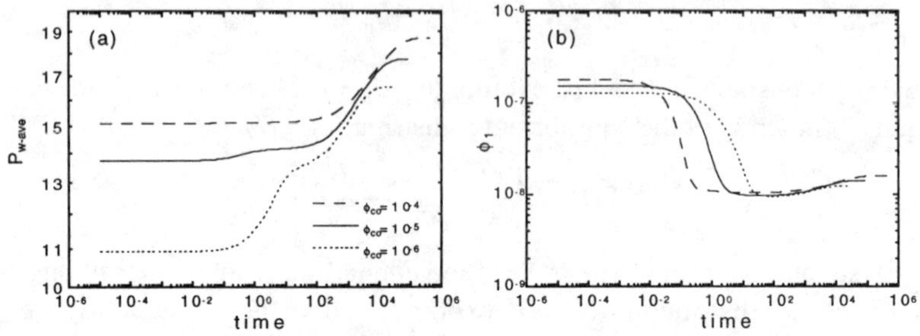

Fig. 6. Results of numerical calculations based on the A-W theory for changes of weight-average association number $P_{w\text{-}ave}$ and unimer concentration ϕ_1 with time in the micellar relaxation of shallow-to-deep quench.

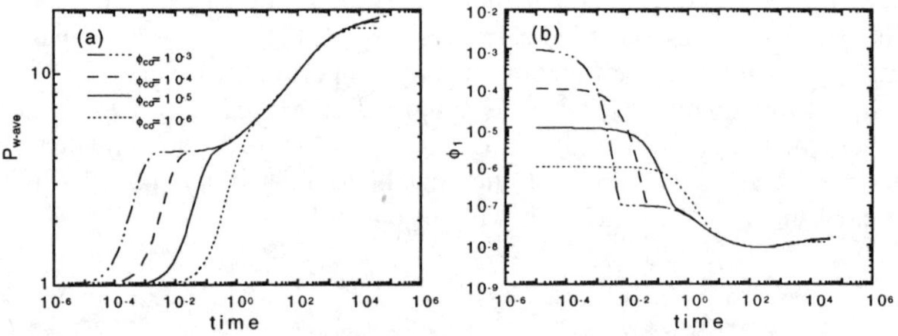

Fig. 7. Results of numerical calculations based on the A-W theory for changes of weight-average association number $P_{w\text{-}ave}$ and unimer concentration ϕ_1 with time in the micellization from unimer state.

The phased growth of micelles can more distinctly be observed in the micellization than the micellar relaxation, in particular in time evolution of the weight-average association number $P_{w\text{-}ave}$. The results of present calculations are consistent with experimental observations of the two-step growth in $P_{w\text{-}ave}$ (Figure 1a) and concentration dependence of the rates (Figure 1b). According to the present theory, the first step of micelle formation experimentally observed may involve the fast process of micellar relaxation as well as the very fast process of micelle formation.

Detailed kinetic behaviors surely depend on the temperature depth from critical micelle temperature and temperature history including the distance from the equilibrium as well as the block composition, the degree of polymerization and the copolymer concentration, although the characteristic features mentioned above are maintained.

Concluding Remarks

The process dependence is unavoidable and sometime essential in studying the micelle structure. Strongly-segregated micelles are usually formed under undefined processes. Ideally, one should start with the unimer state, which is a well-defined state, to form micelles through a simple path, and studies on micellization behavior near the onset of micelle formation will be helpful to make clear discussion. Even in such cases, the most difficulty arises from the fact that it usually takes an extremely long time (see time scales in Figures 1, 2, and 3, for instance) for micelles to reach the equilibrium size and structure. In view of this, kinetic studies are really needed, and will attract increasing attention.

References

1) Price, C.; Chan, E. K. M.; Hudd, A. L.; Stubbersfield, R. B. *Polymer Commun.* **1986** *27* 196-198
2) Stejskal, J.; Hlavatá, D.; Sikora, A.; Konák, C; Pltil, J.; Kratochvíl, P. *Polymer* **1992** *33* 3675-3685
3) Stejskal, J.; Konák, C; Helmstedt, M.; Kratochvíl, P. *Collect. Czech. Chem. Commun.* **1993** *58* 2282-2289
4) Hlavatá, D.; Stejskal, J.; Pletil, J.; Konák, C; Kratochvíl, P.; Helmstedt, M.; Mio, H.; Laggner, P. *Polymer* **1996** *37* 799-805
5) Yu, K.; Zhang, L.; Eisenberg, A. *Langmuir* **1996** *12* 5980-5984
6) Zhang, L.; Shen, H.; Eisenberg, A. *Macromolecules* **1997** *30* 1001-1011
7) Iyama, K.; Nose, T. *Polymer* **1998** *39* 651-658
8) Iyama, K.; Nose, T. *Macromolecules* **1998** *21* 7356-7364
9) Aniansson, E. A. G. in *Aggregation Processes in Solution* eds. E.Wyn-Jones and J. Gormally. **1983** Elsevier, Amsterdam, p 70-99
10) Aniansson, E. A. G.; Wall, S. N. *J. Phys. Chem.* **1974** *78* 1024-1030
11) Aniansson, E. A. G.; Wall, S. N. *J. Phys. Chem.* **1975** *79* 857-858
12) Honda, C.; Hasegawa, Y.; Hirunuma, R.; Nose, T. *Macromolecules* **1994** *27* 7660-7668
13) Honda, C.; Abe, Y.; Nose, T. *Macromolecules* **1996** *29* 6778-6785
14) Nose, T.; Iyama, K. submitted to *Computational and Theoretical Polym. Sci.*
15) Leibler, L.; Orland, H.; Wheeler, J. C. *J. Chem. Phys.* **1983** *79* 3550-3557
16) Halperin, A. *Europhys. lett.* **1989** *8* 351-356
17) Halperin, A.; Alexander, S. *Macromolecules* **1989** *22* 2403-2412
18) Halperin, A.; Tirrell, M.; Lodge, T. P. *Adv. Polym. Sci.* **1992** *100* 31-71
19) Nitzan, A. *Adv. Chem. Phys.* **1988** *70* 489-555

Amphipathic Synthetic and Biological Macromolecules in Aqueous Media

C.L. McCormick,[1,2] R.S. Armentrout,[1] G.C. Cannon,[2] and G.G. Martin[2]

[1]Department of Polymer Science, [2]Department of Chemistry and Biochemistry
University of Southern Mississippi, Hattiesburg, MS 39406

Introduction

Biological and synthetic amphipathic macromolecules are often capable of forming phase-separated microdomains in response to external stimuli, for example pH, ionic strength, shear stress, and temperature. In this manuscript, we discuss the aqueous solution behavior of two synthetic polymers and a biopolymer that possess microstructural features which allow reversible associations. The first synthetic series was prepared by hydrolyzing an octylamide-modified polymer of maleic anhydride [MA] and ethyl vinyl ether [EVE]. Responsiveness to pH has been studied via energy transfer utilizing photophysical probes. The second series of copolymers was synthesized by cyclocopolymerization of *N,N*-diallyl-*N,N*-dimethyl ammonium chloride [DADMAC] and 3-(N,N-diallyl-N-methylammonio)-1-propanesulfonate [DAMAPS]. Electrolyte responsiveness of these novel sulfobetaine copolymers is discussed as a function of composition. The final amphipathic water-soluble polymer is hydrophobin, a fungal protein secreted by *Schizophyllum commune*. This protein is capable of self-assembly from water onto both hydrophilic and hydrophobic surfaces as demonstrated by contact angle and AFM studies.

In each of these architectural types, stimuli-responsiveness of the microstructural elements allows conformational changes resulting in macroscopic ordering into associative domains. Elucidation of the precise nature of these interactions should prove of great utility in preparing systems with capture/delivery capability for pharmaceutical or environmental applications.

Experimental

N-Octyl-substituted copoly(sodium maleate-alt-ethyl vinyl ethers) [poly(SM-EVEs)] were prepared with energy transfer labels utilizing previously reported procedures [1,2]. The M_w values of the 30 mol% octyl substituted polymers were found to be 3.5×10^5 by classical light scattering. Substitutional compositions of the singly-labeled and doubly-labeled dansyl and napthyl polymers [Table 1] were determined by ^1H and gated decoupled ^{13}C NMR (D_2O) and UV absorbance data.

Y. Morishima · T. Norisuye · K. Tashiro (Eds.)
Molecular Interactions and Time-Space Organization in Macromolecular Systems
Proceedings of the OUMS '98, Osaka, Japan, 3-6 June, 1998

Copolymers of 3-(*N,N*-Diallyl-*N*-methylammonio)-1-propanesulfonate (DAMAPS) with *N,N*-diallyl-*N,N*-dimethyl ammonium chloride [DADMAC] and the respective homopolymers were synthesized by free radical photopolymerization in 0.5 *M* aqueous NaCl solutions under nitrogen at 35°C utilizing the photoinitiator Irgacure 2959 [3,4]. Molecular weights, determined by classical light scattering in aqueous 1 *M* NaCl solutions, ranged from $3x10^4$ to $6x10^4$. Compositions were determined by gated-decoupled ^{13}C NMR [Table 2].

Hydrophobin, a 24kD protein, was obtained by a purification procedure [5,6] of one liter standing culture produced by the filamentous fungus *Schizophyllum commune*. Molecular weight was confirmed by SDS-PAGE analysis. Water contact angle measurements and atomic force microscopy (AFM) measurements were performed on coated surfaces as delivered from aqueous solution.

Results and Discussion

Hydrophobically Modified Poly(SM-EVEs)

N-octylamide-substituted poly(sodium maleate-alt-ethyl vinyl ethers) are representative of a group of polyelectrolytes which possess a random coil-to-globule transition in aqueous solution. Such systems, termed "polysoaps" by Straus and coworkers in the 1950's[7] form micelle-like microdomains in aqueous media. The extent and reversibility of domain association depend upon a number of intrinsic factors including microstructural composition and arrangement, charge distribution, electrolyte concentration, pH, temperature, and molecular weight.

Originally, "polysoaps" were thought to mimic the behavior of singular globular proteins (unimers) in aqueous media – possessing largely intramolecular associations along the backbone. However, the interiors of most "polysoaps" are much more loosely packed than native proteins. In fact, most synthetic polymers also exhibit intermolecular modes of association which become apparent only above the critical overlap concentration, C* [8,9].

In this section, we illustrate a) the pH dependence on formation of intramolecular microdomains (often termed micelles) and b) the concentration at which intermolecular associations occur. This is accomplished by non-radiative energy transfer (NRET) studies on random 30 mol% octylamide-modified (SM-EVEs) singly and doubly labeled with napthyl or dansyl labels. Table 1 illustrates the composition for the synthesized copolymers. Note that two spacer lengths were utilized for both napthyl (N1 and N8) and dansyl (D2 and D8) donor/acceptor pairs.

For energy transfer to occur, the donor and acceptor chromophores must be within the Förster distance, previously determined by be 23.45Å for the napthyl/dansyl pair [10]. In our pH studies, we have maintained a total polymer concentration of 0.50 g/L and labeled chromophore concentrations of $1.0x10^{-5}$*M*. NRET quantum efficiency, χ, has been calculated at each pH value as previously reported by our group [1,2] utilizing a modified method reported by Guillet [11].

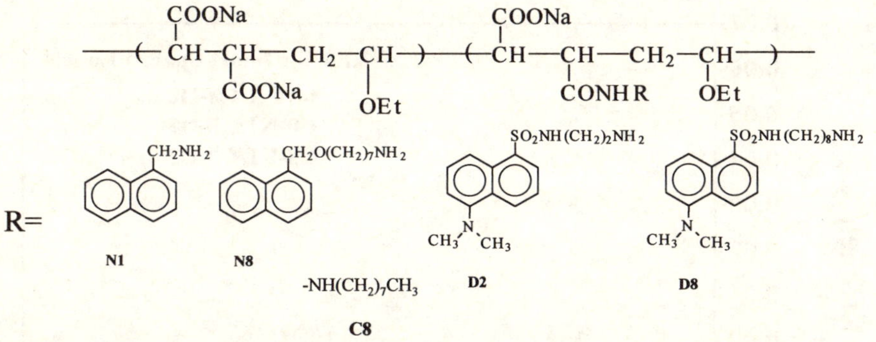

Scheme 1. Hydrophobically-modified copoly(sodium maleate-*alt*-ethyl vinyl ethers) with fluorescent labels.

Table 1. Composition of N-octylamide modified and unmodified copoly(sodium maleate-*alt*-ethyl vinyl ethers)

Polymer	C8 Content (Mol. %)[a]	Naphthalene Content (Mol. %)[b]	Dansyl Content (Mol. %)[b]
C8	30	-	-
C8-N1	31	1.0	-
C8-D2	30	-	1.0
C8-N1-D2	30	1.2	1.0
C8-N8	32	1.1	-
C8-D8	30	-	1.0
C8-N8-D8	30	1.1	1.0
C0	-	-	-
C0-N1	-	1.1	-
C0-D2	-	-	1.0
C0-N1-N2	-	1.0	1.0
C0-N8	-	1.1	-
C0-D8	-	-	1.0
C0-N8-D8	-	1.1	1.0

[a]From ^{1}H and gated-decoupled ^{13}C NMR (D_2O) data [b]From UV absorption data

Figure 1 illustrates results of experiments in which separate napthyl- and dansyl-labeled chains were mixed followed by pH adjustment. As expected, the energy transfer efficiency for the controls (C0-N1/C0-D2) and (C0-N8/C0-D8) with no octyl modification as well as for the octyl-substituted derivatives in 6*M* urea (a breaker of hydrophobic domains) was nearly zero at all pH values. Somewhat surprising, however, were the small, yet experimentally significant values, observed even at high pH for chromophores possessing both the long and short spacers. NRET increases four-fold for the former and two-fold for the latter on lowering the pH to 4.0. Clearly, intermolecular associations are occurring upon domain formation even at these low polymer concentrations.

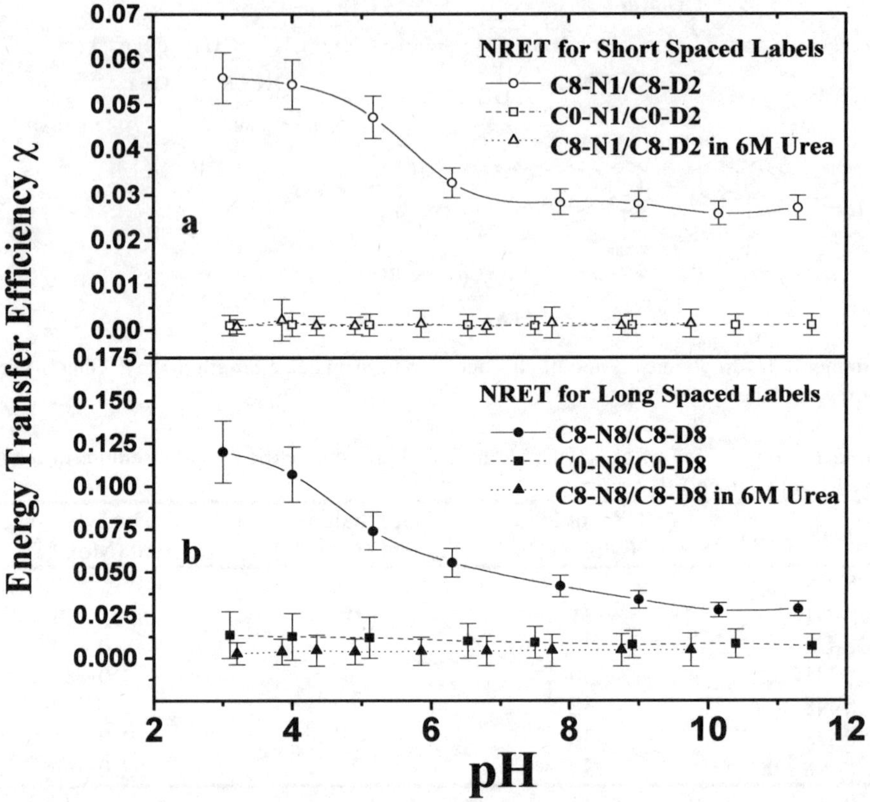

Figure 1. Plot of NRET quantum efficiency for mixtures of singly-labeled copolymers as a function of pH at a concentration of 0.50 g/L for (a) short-spaced labels and (b) long-spaced labels.

Figure 2 shows energy transfer as a function of pH for doubly labeled systems. For the hydrophobically-modified (C8) systems, the polysoap-to-polyelectrolyte transition is clearly illustrated by the S-shaped curve with pH as the randomly (double) labeled domains are disrupted. The NRET efficiency of the C8/C8-N1-D2 polymer is approximately twice as high as that of the unmodified C0/N1-D2-C0 polymer over the entire pH range studied. The longer spacer lengths in the C8/C8-N8-D8 polymer allow efficient accessibility to the hydrophobic octyl domains especially below pH 7.0. Urea disrupts interactions across the entire pH range.

Close study of Figures 1 and 2 leads to the following conclusions: a) hydrophobic microdomains increase as pH is lowered; b) hydrogen bonding at low pH enhances intermolecular interactions, c) multi-molecular (multimer) associations can occur, even in dilute solutions. Importantly, the classical unimolecular micelle concept for polysoaps is not all inclusive and should be reconsidered.

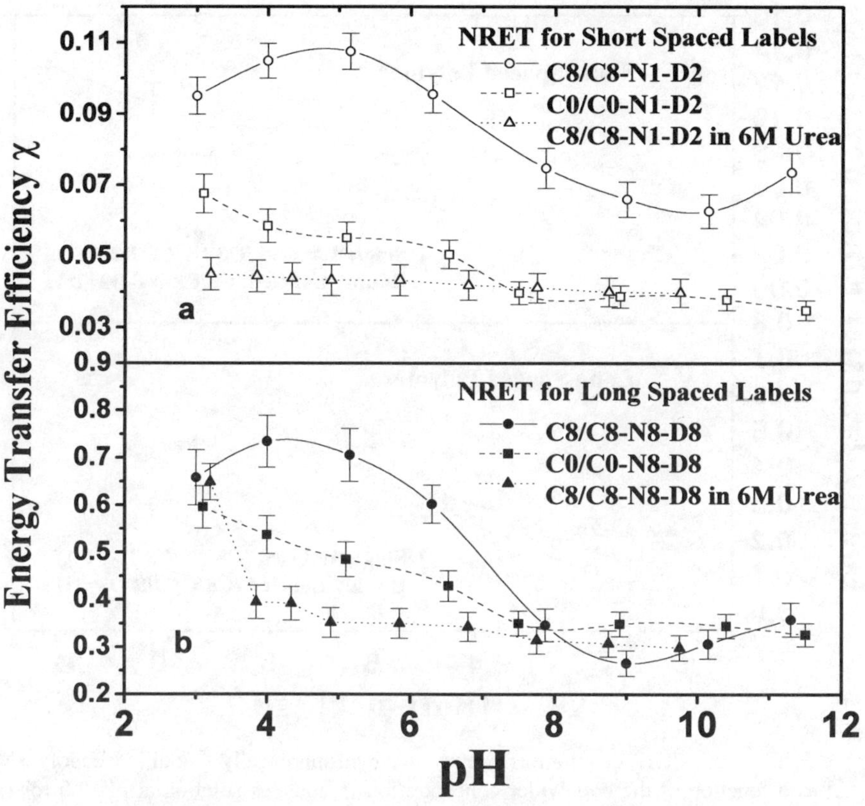

Figure 2. Plot of NRET quantum efficiency for mixtures of doubly-labeled copolymers as a function of pH at a concentration of 0.50 g/L for (a) short-spaced labels and (b) long spaced labels.

Figure 3 illustrates the expected consequence of increasing polymer concentration at a single pH value (9.0) for labeled octylamide-modified copoly(SM-EVEs). The intersection points of the respective curves for the singly and doubly labeled polymers with short (panel a) and long (panel b) spacers represent critical concentrations (11 g/L and 0.7 g/L, respectively) at which statistical interactions of donors and acceptors become equivalent in aqueous solution. This may be a microscopic representation of the critical overlap concentration, C*, traditionally described for the transition from the dilute to the semidilute regime. Additional information of pH and electrolyte responsiveness of octylamide-modified copoly(SM-EVEs) in aqueous media including domain size, aggregation number, and dynamic behavior has been gained by fluorescence quenching, time resolved fluorescence anisotropy, and dynamic light scattering [1,2,12].

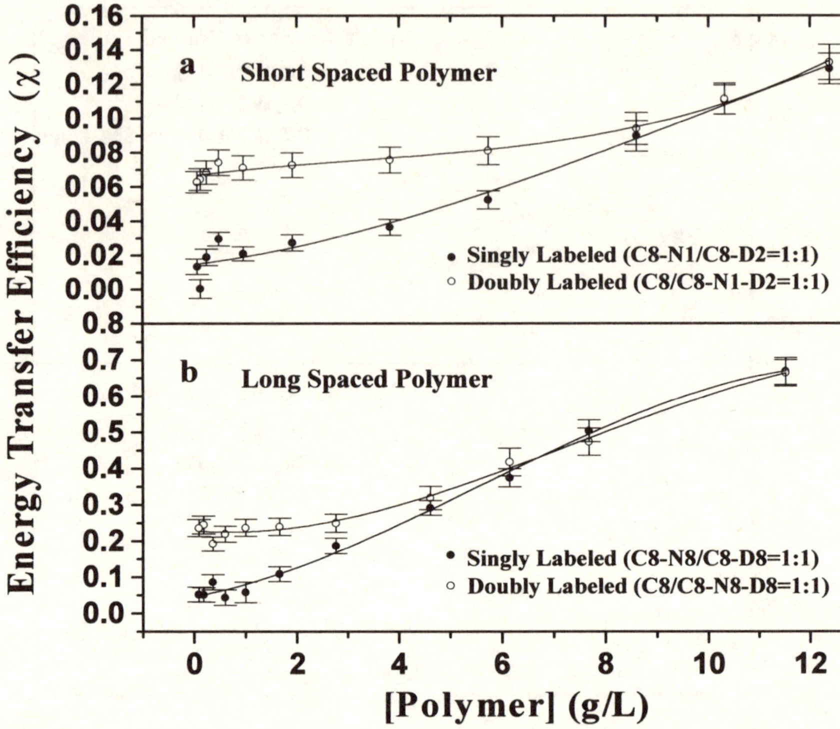

Figure 3. Plots of NRET quantum efficiency of hydrophobically modified copoly(SM-EVEs) as a function of the copolymer concentration in aqueous solution at pH 9.0 for (a) short- and (b) long-spaced labels.

Zwitterionic Cyclopolymers with Sulfobetaine Units

Polyampholytes or polymers possessing both anionic and cationic units pendent to the polymer backbone have interesting behavior in aqueous solution. Hydrophilic polyzwitterionic polymers prepared by incorporation of equimolar quantities of cationic and anionic monomers or with a zwitterionic comonomer, exhibit increases in viscosity upon addition of electrolyte. This "anti-polyelectrolyte" behavior has been attributed to disruption of intramolecular (dipolar) interactions.

Sulfobetaine contatining copolymers have been extensiely studied [13,14,15]. Most systems are based on styrenyl, pyridinyl, acrylic, or acrylamido units. In this contribution we introduce cyclopolymers containing both zwitterionic sulfobetaine units and cationic units [3,4]. Addition of electrolyte to aqueous solutions affects solvation of the cationic groups differently than the zwitterionic groups.

First, we prepared the monomer, 3-(*N,N*-Diallyl-*N*-methylammonio)-1-propanesulfonate (DAMAPS) by the ring opening reaction of 1,3-propane sultone with *N,N*-diallyl-*N*-methylamine. Cyclopolymers of DAMAPS and DADMAC

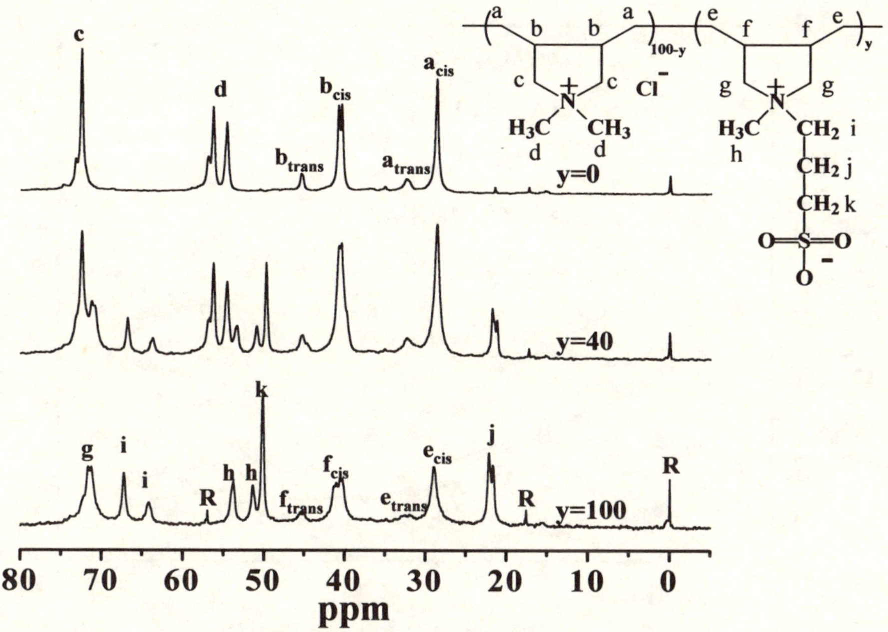

Figure 4. Representative gated decoupled ^{13}C NMR spectra of the DADMAC-co-DAMAPS cyclocopolymers. R=3-(trimethylsilyl)-1-propanesulfonic acid, sodium salt as a reference.

Table 2. Reaction parameters, copolymer composition, and molecular weight for the copolymerization of DADMAC (M_1) and DAMAPS (M_2).

Sample	M_2 Feed (mol%)	M_2 Content (mol%)[a]	*cis* Isomer (mol%)[a]	M_w x 10^{-5} (g mol^{-1})[b]
DADMAC	0	0[c]	89	3.04
DADS11	10	11	90	3.38
DADS23	25	23	90	4.28
DADS40	40	40	89	5.44
DADS55	60	55	88	4.97
DADS82	80	82	89	5.52
DAMAPS	100	100[c]	89	6.03

[a] From gated-decoupled ^{13}C NMR (D_2O) [b]Multi-angle light scattering in 1 *M* NaCl
[c]Theoretical value

and their respective cyclocopolymers were prepared in 0.5 *M* NaCl utilizing a photoinitiator. Table 2 lists initial feed compositions, copolymer compositions, and molecular weight data. Copolymer compositions were determined via gated decoupled ^{13}C NMR. Representative structures and spectral assignments are given in Figure 4. Assignments of the five-membered, homopolymer structure of DADMAC (y=0) [16] were helpful in making the final designations for the sulfobetaine homopolymer (y=100) and the respective copolymers.

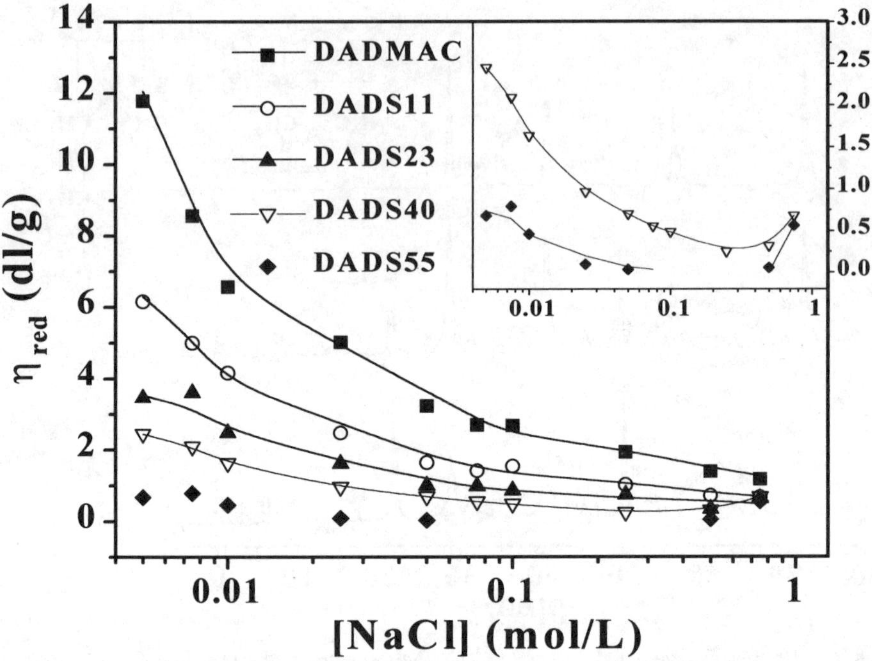

Figure 5. Reduced viscosity for the DADS series at 25°C as a function of electrolyte concentration ([Polymer] = 0.15 g/dl, $\gamma = 5.96\ s^{-1}$).

Quite interesting solution behavior is observed for this series (DADS) with addition of NaCl. For polymers with less than 40 mol% DAMAPS incorporation, typical polyelectrolyte behavior is observed – that is a reduction in the reduced viscosity with an increase in NaCl concentration. This response is lessened as the sulfobetaine content increases. For cyclopolymers with >40 mol% DAMAPS incorporation, an increase in reduced viscosity is observed at high NaCl concentrations due to a screening of the dipole-dipole attractions ("antipolyelectrolyte effect"). Interestingly, a critical point is reached for the DADS55 polymer in which spontaneous separation into polymer rich and polymer poor phases occurs. This is best illustrated in the insert of reduced viscosity versus NaCl concentration [Figure 5]. Single phases can be attained by increasing or decreasing the electrolyte content as evidenced by an increase in reduced viscosity. Optical microscopy demonstrates the effect of increasing electrolyte from 0.1 *M* to 0.5 *M* at a polymer concentration of 1.0 g/dl [Figure 6]. These results indicate that the balance of hydrophobic and hydrophilic forces can be changed simply by changing the solvation of the respective cationic and zwitterionic units.

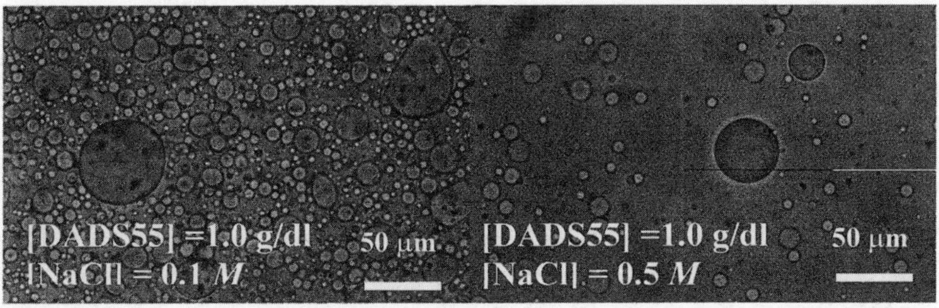

Figure 6. Optical micrograph of aqueous solutions of DADS55 in 0.1 and 0.5 *M* NaCl.

Hydrophobin, An Amphipathic Protein

Perhaps one of the most intriguing water-soluble polymers studied in our laboratories is the 24kD protein, hydrophobin, produced by the filamentous fungus, *Schizophyllum commune*. Although the precise biological function of this protein is still unknown, it is thought to play important roles in the production and emergence of aerial hyphae and in fungal attachment to the host organism. Extensive research by Wessels and Wösten has demonstrated the importance of amphipathic nature of the protein for ordering onto both hydrophobic and hydrophilic surfaces [17,18,19].

In our work, we have observed that although hydrophobin alone can adsorb strongly onto hydrophobic surfaces, its transfer to surfaces is greatly enhanced by a high molecular weight 1→3 linked polysaccharide, schizophyllan, which is coproduced by the fungus in aqueous growth media [5,6].

The synergistic interactions of the polysaccharide and hydrophobin appear to prevent hydrophobin aggregation. A protective colloid-like mechanism can be imagined in which the hydrophobin is dispersed (emulsified), being sterically stabilized by the polysaccharide. Glycosolated residues of the protein may also play a significant role.

In order to elucidate the mechanism of molecular assembly from water onto hydrophilic and hydrophobic surfaces, we conducted experiments with the culture supernatant. First, these solutions containing hydrophobin and schizophyllan were coated onto mica (hydrophilic) and parafilm (hydrophobic) surfaces. Contact angles were then measured prior to and after washing with hot sodium dodecylsulfate. Atomic force micrographs were obtained as well for hydrophobin on the dried surfaces. Culture supernatant solutions were then treated separately with proteinase K, sodium periodate, and a glucosidase with the intent of degrading the hydrophobin, glycosolated sites on the protein, and the polysaccharide, respectively. Again, mica and parafilm surfaces were coated with each solution and surface contact angle and AFM studies conducted before and after hot sodium dodecylsulfate washes. Results of water contact angle and AFM measurements on mica and parafilm are given in Figures 7, 8, and 9, respectively.

Taken in concert, water contact angle and AFM reveal the importance of the 24kD protein and the schizophyllan polysaccharide on the extent of surface association and integrity of the coated surfaces. Unaltered culture supernatant is

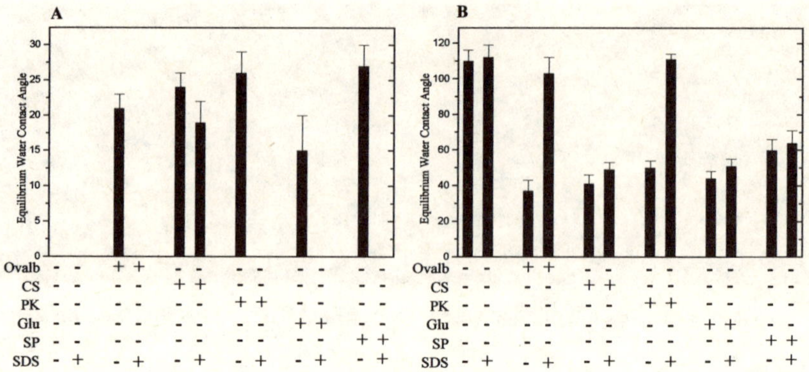

Figure 7. Equilibrium water contact angles. **Panel A:** Mica; **Panel B:** Parafilm®. Ovalb: Ovalbumin; CS: Amicon-concentrated fungal culture supernatant; PK: proteinase K-treated culture supernatant; Glu: β-glucosidase-treated culture supernatant; SP: sodium periodate-treated culture supernatant; SDS: hot sodium dodecyl sulfate wash after coating the surface.

easily coated onto parafilm or mica surfaces resulting in coalescence and apparent uniform surface coverage. The contact angles indicate binding of the amphipathic protein in such a manner that the surface coated mica is rendered more hydrophobic and the parafilm more hydrophilic than prior to treatment. Proteinase K treated supernatant in essence destroys the binding resin and residual components are readily removed by SDS. Treatment of supernatant with B-glucosidase apparently prevents stabilization of the hydrophobin by degrading the protective polysaccharide, schizophyllan. No binding of hydrophobin is detected by contact angle or AFM. Similar behavior is observed when the mannose units (glycosylated sites on hydrophobin) are removed.

Results of coating onto parafilm are considerably different. Contact angle and AFM data indicate that removal of schizophyllan or mannose residues does not prevent substrate coating. Contact angles are similar before and after hot SDS washing. AFM images show coalescence of hydrophobin onto parafilm, although the extent of coverage appears to be different after SDS treatment. This seems reasonable since upon removal of either mannose units or the protective colloid, schizophyllan, self-aggregation could compete with surface adsorption. Some surface features before SDS washing suggest that self-assembled hydrophobin aggregates are easily washed away, yet enough remains on the surface to change the contact angle.

We suggest a model (Figure 10) consistent with the mechanism of steric stabilization of colloids in aqueous media. Hydrophobin in supernatant culture exists as discrete, dispersed particles stabilized by the protective polysaccharide schizophyllan. The stabilized entities can be deposited onto hydrophilic or hydrophobic surfaces, pathways A or B. The chief role of the schizophyllan, perhaps in concert with the mannose residues, is to prevent self-aggregation. Enzymatic degradation of the hydrophilic protective layer does not prevent deposition (and coalescence) onto a hydrophobic surface but allows competitive self-coalescence to rod-like entities. Deposition onto hydrophilic surfaces of hydrophobin stripped of its protective layer, however, does not occur.

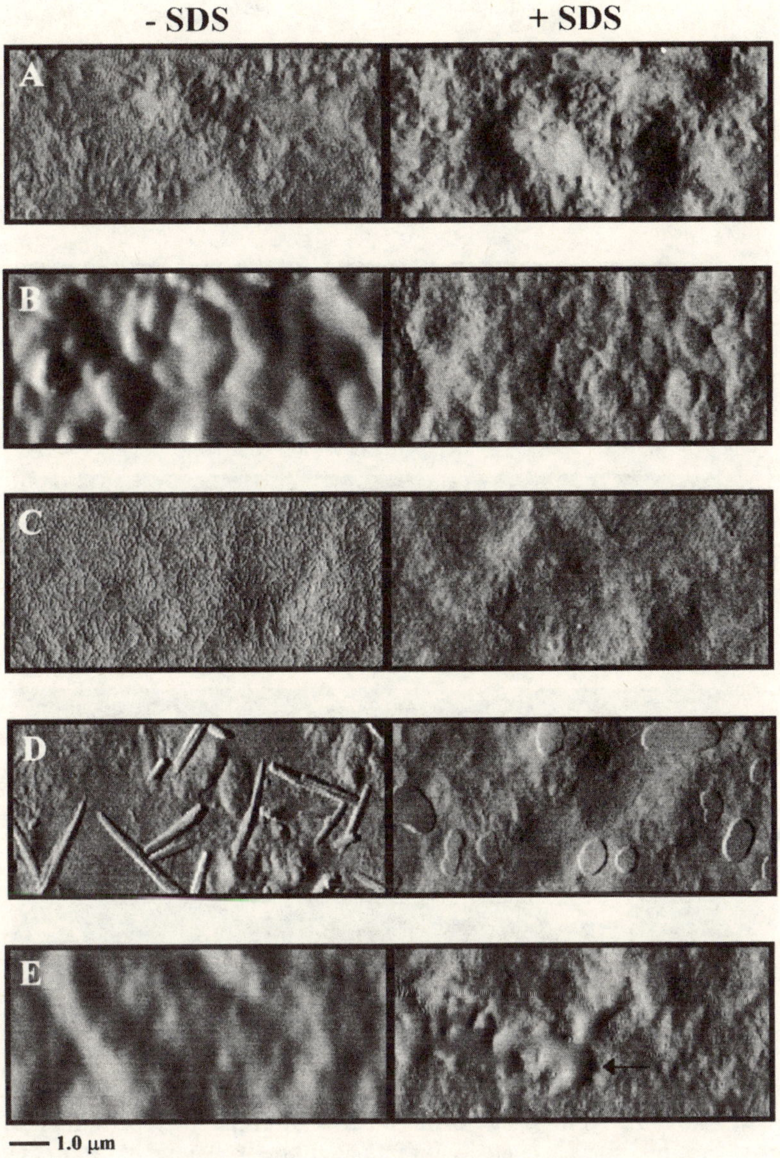

Figure 8. Panel A: AFM image of uncoated mica. **Panel B:** AFM image of mica surface coated with fungal culture supernatant. **Panel C:** AFM image of mica surface coated with proteinase K-treated culture supernatant. **Panel D:** AFM image of mica surface coated with β-glucosidase-treated culture supernatant. **Panel E:** AFM image of mica surface coated with sodium periodate-treated culture supernatant. The left-hand panel shows the surface prior to SDS treatment; the right-hand panel shows the surface after exposure to hot SDS as described in the experimental section. Each panel represents approximately 10.0 μm along the x axis and approximately 4.5 μm along the y axis.

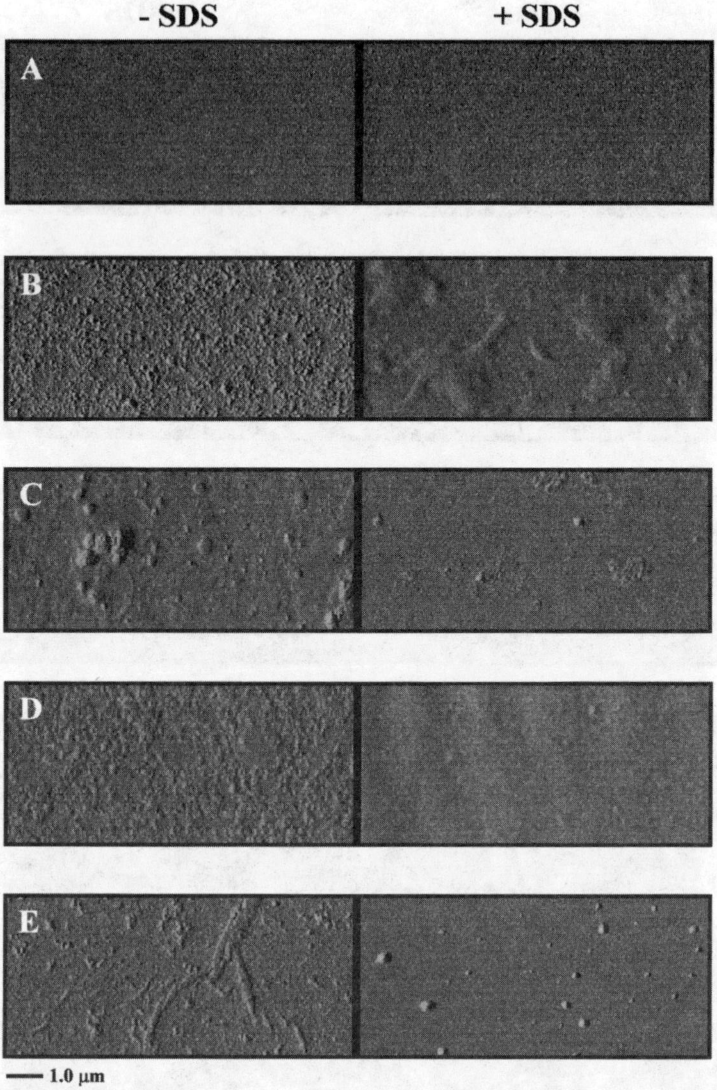

Figure 9. Panel A: AFM image of uncoated Parafilm® surface. **Panel B:** AFM image of Parafilm® surface coated with fungal culture supernatant. **Panel C:** AFM image of Parafilm® surface coated with proteinase K-treated culture supernatant. **Panel D:** AFM image of Parafilm® surface coated with β-glucosidase-treated culture supernatant. **Panel E:** AFM image of Parafilm® surface coated with sodium periodate-treated culture supernatant. The left-hand panel shows the surface prior to SDS treatment; the right-hand panel shows the surface after exposure to hot SDS as described in the experimental section. Each panel represents approximately 10.0 μm along the x axis and approximately 4.5 μm along the y axis.

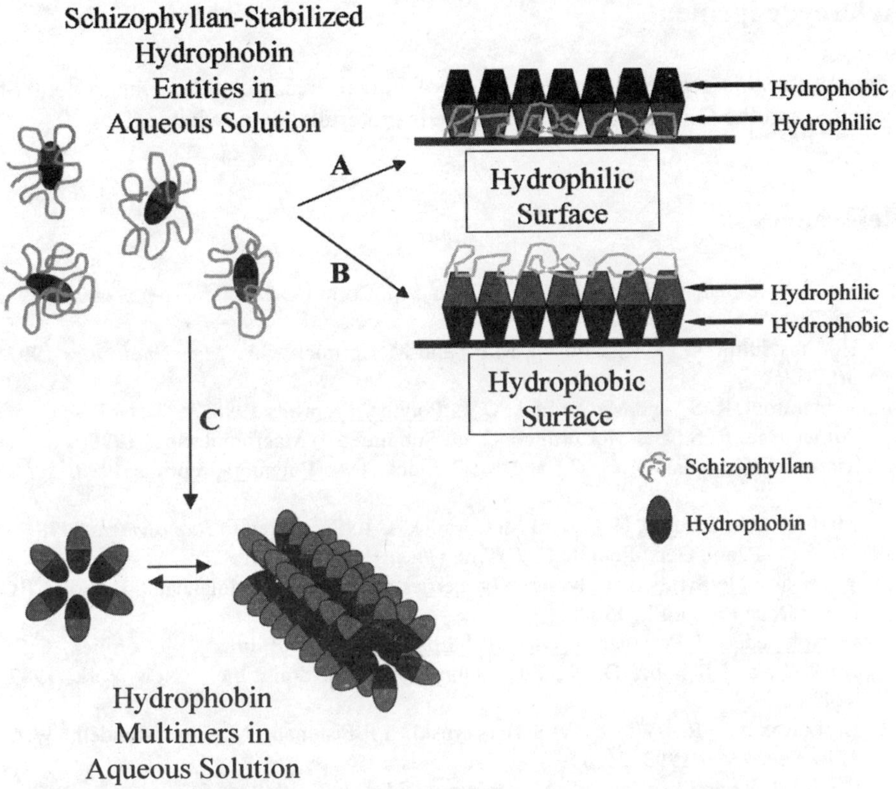

Figure 10. Model of hydrophobin assembly onto hydrophobic or hydrophilic surfaces.

Conclusions

Stimuli responsiveness of three amphipathic polymers: N-octylamide poly[SM-EVEs], zwitterionic cyclocopolymers of DADMAC and DAMAPS, and the fungal protein hydrophobin in aqueous media are representative of the wide range of solution behaviors attainable in aqueous solution. Although rapid advances have been made in recent years in our laboratories and those of Morishima [20,21], F. Winnik [22,23], Candau [24,25], Laschewsky [26,27], and M. Winnik [28,29], to name a few, in understanding the nature of domain formation and stimuli responsiveness in aqueous media, many challenges lie ahead. The early, insightful work of Strauss, Morawetz, and Flory that pointed toward the potential biomimicry in design of "intelligent materials" is rapicly being realized in both academic and insdustrial research laboratories.

Acknowledgements

Support for this research by the Office of Naval Research, the Department of Defense, and the Gillette Research Institute is gratefully acknowledged.

References

1. Hu, Y.; Kramer, M.C.; Boudreaux, C.J.; and McCormick, C.L. *Macromolecules* **1995**, *28*, 7100.
2. Hu, Y.; Smith, G.L.; Richardson, M.F.; and McCormick, C.L. *Macromolecules* **1997**, *30*, 3526.
3. Armentrout, R. S.; and McCormick, C. L. Polymer Preprints **1998**, *39(1)*, 617.
4. Armentrout, R. S.; and McCormick, C. L. Submitted to Macromolecules **1998**.
5. Martin, G.G.; Cannon, G. C.; and McCormick, C.L. Polymer Preprints **1998**, *39(1)*, 347.
6. Martin, G.G.; Cannon, G. C.; and McCormick, C.L. Submitted to *Biopolymers* **1998**.
7. Straus, U.P. and Gershfield, N.L. *J. Phys. Chem* **1954**, *58*, 747.
8. Water-Soluble Synthetic Polymers: Properties and Behavior, Moleneaux, P., Ed.; CRC Press: Boca Raton, FL, **1984.**
9. Encyclopedia of Polymer Science and Engineering, 2nd Edition; McCormick, C.L.; Bock, J.; and Schultz, D. N., Eds.; John Wiley and Sons, Inc.; New York, **1989**; Volume 17.
10. Lackowicz, J. R.; Wiczk, W.; Gryczynski, I.; Fishman, M.; and Rendell, W.A. *Macromolecules* **1993**, *26*, 349.
11. Guillet, J. E.; and Rendell, W. A. *Macromolecules* **1986**, 19, 224.
12. Hu, Y.; Armentrout, R. S.; and McCormick, C.L. *Macromolecules* **1997**, *30*, 3538.
13. Bekturov, E.A; Kudarbergenov, Rafikov, S.R. *J. Macromol. Sci-Rev. Macromol. Chem. Phys*. **1990**, *C30(2)*, 223.
14. Salamone, J.C.; Rice, W.C. *Encyclopedia of Polym. Scie. and Eng*., 2nd Edition; Wiley-Interscience: New York, **1988**, *11*, 514.
15. McCormick, C.L.; Kathmann, E.E. Polymeric Materials Encyclopedia; CRC Press: Boca Raton, **1996**, *7*, 5462.
16. Lancaster, J.E., Baccei, L., Panzer, H.P. *Polymer Letters*, **1976**, *14*, 549.
17. Wösten, H.A.B.; de Vries, O.M.H.; and Wessels, J.G.H. *Plant Cell.* **1993**, *5*, 1567.
18. Wösten, H.A.B.; Ásgeirsdóttir, S.A.; Krook, J.H.; Drenth, J.H.H.; and Wessels, J.G.H. *Eur. J. Cell Biol.* **1994**, *63*, 122.
19. Van der Vegt, W.; van der Mei, H.C.; Wösten, H.A.B.; Wessels, J.G.H.; and Busscher, H.J.,*Biophys. Chem.* **1996**, *57*, 253.
20. Yamamoto, H., Mizusaki, M., Yoda, K., Morishima, Y. *Macromolecules* **1998**, *31*, 3588.
21. Morishima, Y., Nomura, S., Ikeda, T., Seki, M., Kamachi, M. *Macromolecules* **1995**, *28*, 2874.
22. Yamazaki, A., Song, J. M., Winnik, F.M, Brash, J.L. *Macromolecules* **1998**, *31*, 109.
23. Winnik, F. M., Regismond, S., Goddard, E.D. *Langmuir* **1997**, *13*, 111.
24. Stahler, K; Selb, J; Barthelemy, P; Pucci, B; Candau, F. *Langmuir* **1998**, *14*, 4765.
25. Pabon, M., Selb, J., Candau, F. *Langmuir* **1998**, *14*, 735.

26. Delcorte, A; Bertrand, P; Wischerhoff, E; Laschewsky, A. *Langmuir* **1997**, *13*, 5125.
27. Cochin, D; Hendlinger, P; Laschewsky, A. *Colloid and Polymer Science* **1995**, *273*, 1138.
28. Tam, K C; Jenkins, R D; Winnik, M A; Bassett, D. *Macromolecules* **1998**, *31*, 4149.
29. Rager, T., Meyer, W.H., Wegner, G., Winnik, M.A. *Macromolecules* **1997**, *30*, 4911.

The Interactions of Hydrophobically-Modified Poly-(N-Isopropylacrylamides) with Liposomes

Françoise M. Winnik and Alla Polozova
Department of Chemistry, McMaster University, 1280 Main St W Hamilton ON Canada L8S 4M1
Winnikf@mcmaster.ca

Abstract.

The interactions between a random copolymer of N-isopropylacrylamide (NIPAM), N-[4-(1-pyrenyl)butyl]-N-n-octadecylacrylamide and N-glycyl-acrylamide (Gly) (PNIPAM-Py-Gly) with a series of vesicles have been examined by fluorescence spectroscopy. The copolymer exhibits pH-sensitivity due to the carboxylic acid groups of the glycine residues (15 mol %), thermosensitivity due to the NIPAM residues (84 mol %), and amphiphilic characteristics due to the pyrenyl octadecyl moieties (1 mol %). In the presence of salts, complex formation was detected between the copolymer and cationic or anionic non-phospholipid liposomes based on the non-ionic surfactant, n-octadecyl-diethylene oxide ($(EO)_2C_{18}H_{37}$), and cationic phospholipid liposomes. Weak interaction was detected between the copolymer and neutral phospholipid liposomes under the same conditions. The binding of the copolymer to the liposomes is shown to be controlled primarily through hydrogen bond formation between the hydroxyl groups of the $(EO)_2C_{18}H_{37}$ and the amide groups of the NIPAM residues, electrostatic interactions between the charged surfactants of the bilayer and the glycine residues and to, a lesser extent, by hydrophobic forces.

1. Introduction

Liposomes form spontaneously when lipids are dispersed in aqueous media. They consist of a bilayer membrane, which captures a pool of water in its center. The practical value of liposomes derives from two unique properties: 1) their ability to entrap water-soluble materials in the internal water reservoir or liposoluble compounds in the lipid bilayer; and 2) their compatibility with natural membranes, making them safe for medical applications. They have attracted considerable interest as delivery systems, such as carriers of drugs to specific targets in the treatment of cancer tumors, in gene therapy, and in various diagnostic tests using fluorescent markers and radiolabels[1]. Liposomes are fragile structures created in water as a result of a delicate balance of interacting forces, which arise when amphiphilic compounds are added to water. In most practical applications liposomes have to be stabilized so that they maintain their integrity in hostile environments [2]. They should also respond to an external stimulus and deliver the entrapped material once they have reached their target [3].

Clearly, key to the success of the liposome technology is the availability of systems with improved mechanical stability in which the dynamic properties of the membranes are preserved. One option, explored in several laboratories, consists of adsorbing securely on the outer surface of the lipid bilayer a polymer, which performs a dual task: it protects the liposome surface and it responds by some physical change to a stimulus such as a change in temperature, pH, or ionic strength. For instance, thermosensitive polymers, such as poly(N-isopropylacrylamide) (PNIPAM), anchored into the lipid bilayer through hydrophobic substituents can either destabilize the liposome bilayer above a specific temperature, or provide increased stability below that temperature. In water hydrophobically-modified PNIPAM's (HM-PNIPAM), like PNIPAM itself, undergo reversible phase transition when heated above a lower critical solution temperature (32 °C, LCST) [4]. Detailed investigations of this phase transition have established that, at the LCST, the PNIPAM chains undergo a collapse from hydrated extended coils to hydrophobic globules, which aggregate and form a separate phase. It is believed that the extended hydrated form of the polymer contributes to the stabilization of the liposome bilayer [5]. Polyelectrolyte/liposome complexes sensitive to pH changes have also been explored extensively. In certain lipid bilayers, polyelectrolytes can promote phase changes, create local defects in the membrane, or cause aggregation or fusion either by themselves or in the presence of added salt. Alternatively, as demonstrated in the elegant work of Tirrell and collaborators, the polyelectrolyte can promote pH-dependent lateral diffusion of charged lipids in the bilayer and induce reversible permeability of the membrane [6]. Particularly effective in promoting such effects are derivatives of poly(2-ethylacrylic acid) anchored into the bilayer through phospholipid residues.

The design of liposomes sensitive to two different stimuli under controlled conditions is a natural extension of these recent achievements. We present here our progress in the design of such complexes based on the thermo-sensitivity of PNIPAM and the pH-sensitivity of carboxylic acid groups. The copolymer employed, PNIPAM-Py-Gly, Figure 1, consists of a PNIPAM chain that carries at random approximately 15 mole percent of glycine residues as well as a small number of hydrophobic groups labeled with a fluorescent dye [7].

Fig. 1. Structure of the polymer used in this study

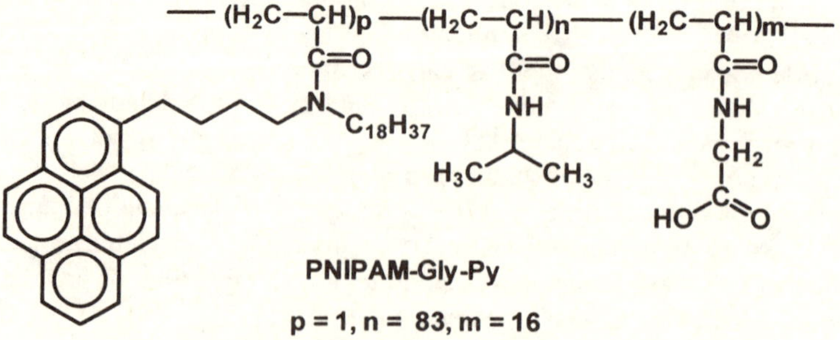

This unique copolymer combines the phase transition characteristics of PNIPAM with the responsiveness to changes in pH, via protonation/deprotonation, of the glycine carboxyl groups. In neutral water, the copolymer forms polymeric micelles, with an average diameter of 20 nm, as a result of inter- and intrapolymeric aggregation of the hydrophobic substituents. The hydrophobic core of such micelles is composed of octadecyl groups and Py labels. It is surrounded by a corona consisting of hydrated poly-(N-isopropylacrylamide) moieties and partially neutralized glycine residues [7].

We report here a study on the interactions of this copolymer with two-types of liposomes: phospholipid liposomes (PLL) and non-phospholipid liposomes (NPL). These vesicles are obtained from mixtures of three types of lipids: (a) a main amphiphilic component, such as n-octadecyldiethylene oxide, (b) a modulator, usually cholesterol, and (c) an ionogen, such as dioctadecyldimethylammonium bromide (DDAB) in the case of cationic NPL's or dioctadecylphosphate (DP) in the case of anionic NPL's. While from a morphological viewpoint the two classes of liposomes bear significant similarity, from the chemical standpoint, the hydrophilic surfaces of the vesicles are quite different. The surface of NPL's is characterized by a high density of hydroxyl and ether functional groups, able to form hydrogen bonds with suitable partners. An intriguing possibility is that of using hydrogen bonds to promote the binding of polymers to liposomes in order to protect the liposome surface or to obtain liposomes that release their content when subjected to specific triggers, such as a rapid change of pH, temperature, or ionic strength.

This concept has been explored in the study described here, using fluorescence assays and by turbidity measurements. The results of this study are reported with particular emphasis on the relative importance of electrostatic forces, hydrophobic interactions, and hydrogen bonds in promoting the formation of liposome/polymer complexes.

2. Experimental

2.1. Materials and methods

Water was purified with a Barnstead NANOPure water purification system. Dimethyldioctadecylammonium bromide (DDAB), dioctadecylphosphate (DP), and palmitoleoyl phosphatidylcholine (POPC) were purchased from Avanti Polar Lipids (Alabaster, Alabama). Cholesterol, n-octadecyldiethylene oxide ($EO_2C_{18}H_{37}$), n-hexadecyldiethylene oxide ($EO_2C_{16}H_{33}$), sodium dodecyl sulfate (SDS) and calcein were obtained from Sigma (Sigma-Aldrich Canada Ltd., Ontario, Canada). BODIPY-PC was supplied by Molecular Probes. Tris, sodium azide (NaN_3), sodium chloride (NaCl), and glycine (Gly) were purchased from BDH Chemicals). A copolymer of N-isopropylacrylamide, N-[4-(1-

pyrenyl)butyl]-N-n-octadecylacrylamide, and N-glycydylacrylamide (PNIPAM-Py-Gly), Mn 25,000, Mw/Mn 2.1 was prepared as previously described [7].

Fluorescence spectra were recorded with a SPEX Fluorolog 212 spectrometer (Edison, NJ) equipped with a DM3000F data system. The temperature of the water-jacketed cell holder was controlled with a Neslab (Portsmouth, NH) circulating bath. The temperature of the sample fluid was measured with a thermocouple immersed in the sample. Emission spectra were recorded with excitation wavelengths of 330 nm (pyrene) or 585 nm (BODIPY). They were not corrected. In experiments aimed at monitoring the adsorption of PNIPAM-Py-Gly on liposomes, the ratio I_E/I_M of the excimer emission intensity to the monomer emission intensity was calculated as the ratio of the emission intensity at 480 nm (excimer) to the half-sum of the emission intensities at 377 nm and 397 nm (monomer). Changes in turbidity (λ = 600 nm) of polymer/liposome mixtures with increasing temperature were determined using a Hewlett Packard 8452A photodiode array spectrometer equipped with a Hewlett Packard 89090A temperature controller and operated under the HP Chemstation Windows-based software. The solutions were heated at a constant rate (0.5 °C min^{-1}) in a magnetically stirred cell. The value reported is the temperature corresponding to the 90 % transmittance point.

2.2. Sample preparation

A solution in chloroform of the desired amounts of lipids was poured into glass test tubes. The solvent was evaporated under a stream of nitrogen. The resulting lipid film was dried under high vacuum for at least two hours. The dry lipid film was hydrated in an aqueous solution of NaCl (150 mM) and NaN_3 (0.02% w/w) to give a lipid suspension of a concentration of 20 g/L. The lipid suspension was subjected to extrusion through polycarbonate filter membranes (200 nm pore size, Nucleopore) using a Lipofast extruder (Avestin, Canada). The hydration and extrusion of the non-phospholipid and POPC liposomes were performed at a temperature of 60 and 25 °C, respectively. Liposome suspensions were added to polymer solutions in the desired proportions. The mixtures were allowed to equilibrate first at room temperature (30 minutes), then at the temperature of the experiment (30 minutes).

3. Results and discussion

3.1. The fluorescence technique.

The fluorescence spectrum of PNIPAM-Py-Gly in water displays two distinct emissions: a broad and strong signal, centered at 480 nm, due to pyrene excimers which originate from chromophores located in close proximity to each other and a well-resolved emission, the pyrene monomer emission, with the (0,0) band at 378 nm, due to locally isolated excited pyrenes (Figure 2) [8]. The excimer emission is very strong in this case, since the hydrophobic pyrene groups are found mostly within the core of the micellar assemblies formed by the polymer in neutral aqueous solution.

The addition of PNIPAM-Py-Gly to a suspension of liposomes triggers remarkable changes in the emission of the labelled polymer. Most noticeable is a sharp increase of the intensity of the monomer emission at the expense of the excimer emission, indicating disruption of the polymeric micelles. The effect is illustrated in Figure 2, which presents spectra of the copolymer in water and in the presence of cationic NPL's. The decrease in excimer emission signals the incorporation of the hydrophobic groups into the lipid bilayer, as the originally tightly packed fluorophores diffuse away from each other within the lipid membrane. We used the change in the ratio I_E/I_M of pyrene excimer to monomer emission intensities as a measure of the degree of polymer binding to liposomes.

Fig. 2. Fluorescence spectra of PNIPAM-Gly-Py in aqueous solution (left) and with cationic liposomes (right); λ_{exc} 346 nm; Polymer concentration: 0.002 g L^{-1}; lipid concentration: 1.0 g L^{-1}; liposome composition: $(EO)_2C_{18}H_{37}$ (75 %): cholesterol (20 %): DDAB (5 %).

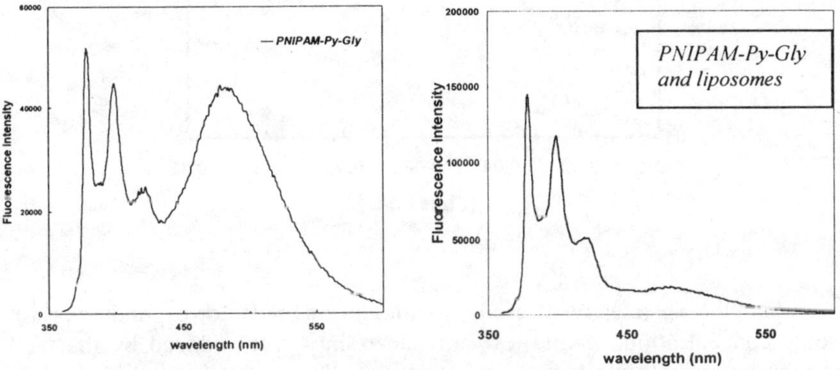

3.2 Interactions of PNIPAM-Gly-Py with cationic liposomes

We examined first the interactions between cationic liposomes and PNIPAM-Py-Gly in neutral solution. In this situation, the liposome/copolymer interactions are expected to be promoted by attractive electrostatic forces. Two types of vesicles were used: (a) non-phospholipid liposomes consisting of the non-ionic surfactant $(EO)_2C_{18}H_{37}$, cholesterol, and the cationic surfactant dioctadecyl-dimethylammonium bromide (DDAB) and (b) POPC liposomes rendered cationic by incorporation of DDAB. The measurements were carried out at constant salt concentration (0.15 M NaCl) in a neutral buffer. Increasing amounts of liposomes were added to a solution of constant polymer concentration (2 mg L^{-1}). This triggered a large decrease of the pyrene excimer emission and a concomitant increase of the pyrene monomer emission, signaling the incorporation of the pyrene groups within the liposome membrane. As more liposomes were added the excimer emission continued to decrease, until it reached a constant value. These changes are recorded in a plot of the changes in the ratio I_E/I_M as a function of lipid concentration presented in Figure 3.

Fig. 3. Change in the ratio I_E/I_M of excimer to monomer emission intensities as a function of lipid concentration in mixtures of PNIPAM-Py-Gly with liposomes of various composition; polymer concentration: 0,002 g L^{-1} in TRIS buffer (5 mM) containing NaCl (0.15 mol L^{-1})

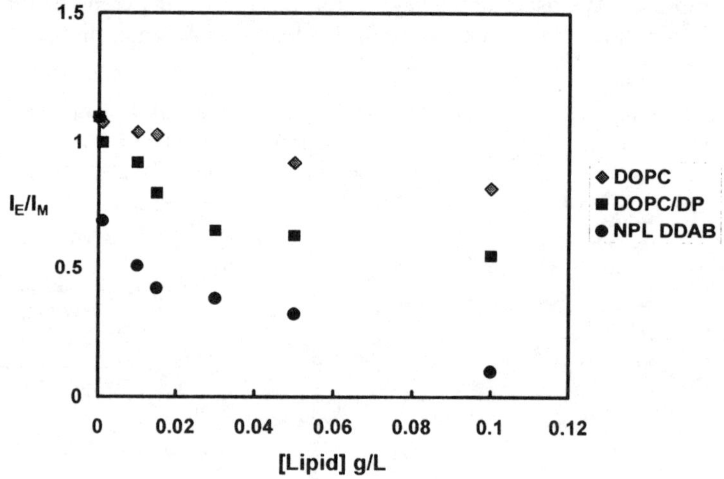

Since unilamellar vesicles were used, the amount of lipids corresponds well to the liposome concentration. Some level of uncertainty is introduced by the fact that there may be a small number of double-layered liposomes. Overall, the value of I_E/I_M dropped from 1.1 in the absence of liposomes to 0.1 after saturation was reached. In the case of cationic NPL's a sharp decrease in excimer intensity upon addition of liposomes was observed up to a polymer/lipid ratio of 1:5 (w/w) (or 8 x 10^{-4} mol/mol). The cationic phospholipid liposomes exhibited the same type of interactions with PNIPAM-Py-Gly, as judged from the decrease in I_E/I_M upon addition of POPC/DDAB liposomes to a solution of PNIPAM-Py-Gly. We note, however, that for similar membrane charge density, provided by 5 mol % of DDAB, the ratio I_E/I_M dropped to a value of 0.6 in the case of the phospholipid liposomes, whereas in the case of NPL's I_E/I_M took a value of 0.1

3.3. Interactions of PNIPAM-Py-Gly with neutral and anionic liposomes

As a control experiment neutral POPC liposomes were added to a solution of PNIPAM-Py-Gly. In this case the emission spectrum of the polymer was hardly affected, implying that in the absence of electrostatic attraction the interactions between phospholipid liposomes and PNIPAM-Py-Gly are weak and that, consequently, a large fraction of the polymer remains in solution in the form of unperturbed micelles. To complete the study we tested as well the behavior of the copolymer in the presence of negatively-charged liposomes. We used NPL's consisting of $(EO)_2C_{18}H_{37}$, cholesterol, and dioctadecylphosphate (DP, 5 mol %). To our surprise, the changes in I_E/I_M of PNIPAM-Py-Gly as a function of added anionic NPL's concentration were identical to those observed upon addition of cationic NPL's (Figure 3). This important finding points to the existence of interactions between the copolymer and these NPL's, overcoming the unfavorable electrostatic interactions anticipated to occur between the phosphate groups on the liposomes and the carboxylate moieties of the copolymer.

3.4. Salts effects on the interactions of PNIPAM-Py-Gly and liposomes

Finally, the fluorescence experiments were performed in the absence of added salt. Increasing amounts of cationic liposomes, either POPC liposomes containing 10 wt % DDAB or NPL's charged with 0.5 wt % DDAB, were added to a PNIPAM-Py-Gly solution in deionized water (2 mg L^{-1}). The changes of the ratio I_E/I_M were monitored as a function of total lipid concentration. The ratio decreased sharply upon addition of liposomes of either composition, reaching a saturation value of 0.1 for a lipid concentration of 0.1 g L^{-1}. The decrease in Py excimer induced by cationic NPL's was very similar to that observed when the measurements were conducted under high salt conditions. In contrast, the spectral changes caused by the addition of POPC liposomes to the polymer solution in water were strikingly different; a much more severe decrease in I_E/I_M occurred, implying a strong association of the polyelectrolyte to the cationic phospholipid liposomes in the absence of salt. This result emphasizes the role of the screening effect of salts in controlling the electrostatic interactions between the cationic liposomes and the copolymer

3.5. Temperature effects

All the experiments reported previously were carried out at 25 °C. Since the ultimate objective of this study is to develop thermosensitive liposomes, it was important to examine in detail the effects of heating/cooling cycles on the properties of the polymer/liposome complexes. Aqueous solutions of PNIPAM-Py-Gly exhibit a pH-dependent cloud point, reflecting the effect of the changes in the level of ionization of the carboxylic acid groups on the hydrophobicity of the copolymer [7]. Binding of the polymer to charged liposomes caused a significant change in the LCST behavior of PNIPAM-Py-Gly. In mixtures of the polymer with cationic liposomes, a temperature-induced increase in turbidity was observed, as seen in typical heating/cooling cycles of aqueous solutions of the copolymer in the absence of liposomes. We established that the onset of turbidity depends on the total lipid concentration and on the relative amount of cationic surfactant incorporated in the bilayer. Specifically, the cloud point decreased as the lipid concentration increased and/or as the concentration of charged surfactant increased. The decrease in cloud point reaches a limit for a 1:5 (w/w) polymer/lipid ratio. Further addition of liposomes, at constant polymer concentration, had no effect on the cloud point of the mixture. We attribute the temperature sensitivity of the polymer/liposome mixtures to a gradual neutralization of the carboxylic acid groups of the polymer by the cationic groups on the liposome surface. This charge neutralization mechanism competes with the pH-sensitive protonation of the polymer, allowing the control of the cloud point via both the pH of the medium and the charge density of the liposome. This property can be exploited to control the leakage of various materials encapsulated into PNIPAM-Py-Gly/liposomes complexes.

4. Conclusions

The solution properties of PNIPAM-Py-Gly reflect its ambivalent chemical composition. It can be viewed as a polyelectrolyte, featuring approximately 130 ionizable carboxylic groups per chain. It is also an amphiphilic polymer, carrying on average 10 hydrophobic octadecyl groups per macromolecule. The interplay between the intrinsic properties of PNIPAM and the specific physicochemical properties of the modifying groups determines the solution properties of the copolymer and its *modus operandi* with liposomes of different compositions under various situations. Adsorption of polyelectrolytes by oppositely charged surfaces occurs when the energy of adsorption exceeds the entropy loss associated with the number of configurations available to the polyelectrolyte. The conditions of polymer adsorption have been treated in several theoretical studies. Of particular relevance to the present work, is the problem of understanding how large scale patterns on surfaces get recognized by macromolecules. This problem has been addressed recently by Muthukumar, in the case of a pattern made up of electrostatic charges and a polymer consisting of opposite charges distributed along the chain backbone [9]. By making the analogy between polymer adsorption and the binding of an electron in a potential well, he concluded that the kinetics of adsorption take place in two distinct steps: a fast binding process without any registry of pattern, followed by a slow process, where the system attempts to attain the fully registered configuration. The adsorption of PNIPAM-Py-Gly onto charged liposomes provides an experimental system that involves such a pattern recognition process. There are three guiding factors involved: (1) electrostatic interactions, and in the case of phospholipid-containing liposomes, these forces are crucial in determining the binding of the polymer to the lipid membrane, but they are of lesser importance in the case of NPL/polymer interactions; (2) hydrogen bond formation, and this is the predominant factor controlling polymer adsorption to NPL's, overtaking even the electrostatic repulsion experienced by the polymer in the presence of anionic liposomes; and (3) hydrophobic interactions driven by the nonpolar side groups of the polymer. The last parameter, while important, appears to be of similar importance for the two types of liposomes.

References

[1] Handbook of Non-Merdical Applications of Liposomes. Barenholz, Y.; Lasic, D.D.; Eds CRC Press. Boca Raton, 1996; Liposome Technology, 2nd Ed. Gregoriadis, G.; Ed.; CRC Press, Boca Raton, 1992.

[2] Monfardini, C.; Veronese, F. M. *Bioconjugate Chem.*, **1998**, *9*, 418.

[3] Gregoriadis, G.; Liposomes as Drug Carriers: Recent Trends and Progress, Wiley & Sons, New York, 1988.

[4] Ringsdorf, H.; Venzmer, J.; Winnik, F. M. *Macromolecules* **1991**, *24*, 1678.

[5] Schild, H. G. *Prog. Polym. Sci.* **1992**, *17*, 163.

[6] Thomas, J. L.; Tirell, D. A. *Acc. Chem. Res.* **1992,** *25*, 336.

[7] Spafford, M.; Polozova, . A.; Winnik, F. M. *Macromolecules* **1998**, in Press.

[8] Winnik, F. M. *Chem. Rev.* **1993,** ***93,*** 587.

[9] von Goeler, F.; Muthukumar, M. *J. Chem. Phys.* **1994,** *100*, 7796 and references therein.

Controlled Association of Macromolecules in Water: Hydrophobized Polymers and Proteins

Kazunari Akiyoshi, Yoshihiro Sasaki, Kenichi Kuroda, Atushi Ueminami and Junzo Sunamoto

Department of Synthetic Chemistry and Biological Chemistry, Graduate School of Engineering, Kyoto University, Yoshida-Hommachi, Sakyo-ku, Kyoto 606-8501, Japan.
E-mail: g54112@sakura.kudpc.kyoto-u.ac.jp

1. Introduction

In living systems, the association of biopolymers such as DNA, polysaccharides and proteins is controlled in a noncovalent manner, such as by electrostatic interaction, hydrogen bonding, and hydrophobic association. The amphiphilicity of biopolymers is an important factor in their self-organization in water. Various polymer amphiphiles such as amphiphilic block copolymers and hydrophobized polymers have been synthesized, and the structure-function relationships of these self-assemblies have been studied.[1-3]. In artificial systems, however, macromolecular association is still difficult to control. The supramolecular assembly of macromolecules is important in various fields for designing new materials. We describe here our recent studies on the self-assembly of hydrophobized polymers where water-soluble polymers are partly substituted by hydrophobic molecules.

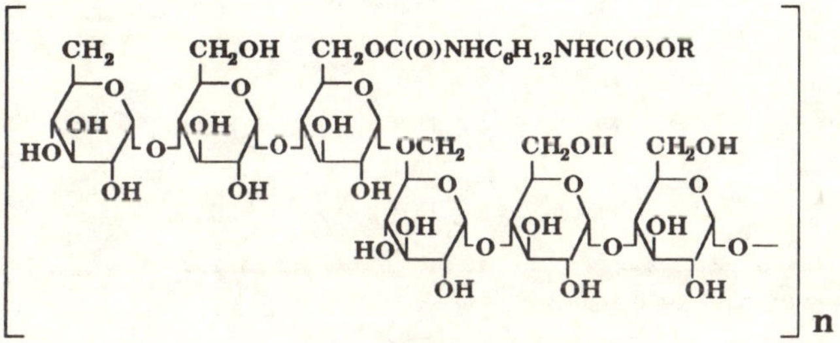

R : -Cholesterol (CHP)
: $-CH(CH_2OC_{12}H_{25})_2$ (2C12P)
: $-C_{12}H_{25}$ (C12P)
: $-C_{16}H_{33}$ (C16P)
: $-C_{20}H_{41}$ (C20P)

Fig. 1. Chemical structure of hydrophobized pullulan

Y. Morishima · T. Norisuye · K. Tashiro (Eds.)
Molecular Interactions and Time-Space Organization in Macromolecular Systems
Proceedings of the OUMS '98, Osaka, Japan, 3-6 June, 1998

. Self-assembly of Hydrophobized Polysaccharides

2.1 Formation of nanoparticles

Water-soluble polysaccharides such as pullulan, dextran and mannan were partly substituted by various hydrophobic groups such as long alkyl chain or cholesterol (Fig. 1). For example, pullulan ($M_w = 55 \times 10^3$, $M_w/M_n = 1.54$) was substituted by 1.7 cholesterol moieties per 100 glucose units of pullulan. Cholesterol-bearing pullulan (CHP) is coded as CHP-55-1.7. The solution properties of hydrophobized polysaccharides were investigated by size-exclusion column chromatography (SEC), light scattering, fluorescence spectroscopy, and electron micrography[4-6].

The sonicated samples of all of the hydrophobized polysaccharides shown in Fig. 1 give a single peak on SEC. The apparent polydispersity of the aggregates was calculated using a calibration curve of standard pullulans, and the M_w/M_n value was below 1.1 for all of the aggregates studied. No precipitation was observed even after heating at 90 °C for 1h. The self-aggregates are very colloidally stable. A transmission electron micrograph showed that the spherical particles have a narrow size distribution. The particle size (hydrodynamic radius, R_H) decreases with an increase in the number of cholesterol moieties of CHP (Table 1).

Table 1. Characteristics of various hydrophobized pullulan

Sample	Aggregation number of polymer[a)]	R_G[a)] (nm)	R_H[b)] (nm)	Φ_G (g/mL)	Aggregation number of hydrophobe
CHP-55-1.1	11	13.5	11.6	0.10	4.4
CHP-55-1.7	10	11.5	10.2	0.15	4.2
CHP-55-2.1	10	11.1	9.5	0.17	5.0
C20P-55-1.6	7.1	11.9	—	0.09	13
C16P-55-2.4	7.4	20.4	—	0.02	15
C12P-55-1.8	8.5	15.6	—	0.05	17
2C12P-55-2.3	13	25.5	—	0.02	11

a) Determined by static light scattering

b) Determined by dynamic light scattering

The molecular weight of the aggregate was determined by SLS. For example, that of an aggregate of CHP-55-1.7 was 5.6×10^5. This suggests that this aggregate consists of approximately ten CHP-55-1.7 molecules. Assuming nondraining hydrodynamic behavior, the average polymer density (Φ_G) of nanoparticles can be calculated from R_H and M_w using the following equation:

$$\Phi_G = \frac{M_W}{N_A} \times \left(\frac{4}{3}\pi R_G{}^3\right)^{-1} \tag{1}$$

where N_A is Avogadro's number. The density of the polysaccharide in one aggregate of CHP-55-1.7 was 0.15 g/mL. Parent pullulan behaves as an expanded flexible coil[7]. For example, the R_G of pullulan with a molecular weight of 5.7×10^5, which is almost the same as that of a CHP self-aggregate, is reported to be 31.5 nm[7]. The Φ_G of pullulan is calculated by eq. 1 to be 0.02 g/mL. The polysaccharide skeleton of CHP must be more densely packed in the CHP self-aggregate than in the parent pullulan. The estimated density of the aggregate is similar to that of agarose or alginic acid hydrogel (0.2-0.4 g/mL). CHP with a higher degree of substitution (DS) of cholesterol forms more densely packed particles due to an increase in the number of cholesterol groups per hydrogel nanoparticle (Table 1). Pullulan with long alkyl chains (ALP) also formed monodispersive nanoparticles by self-aggregation. ALPs form less densely packed nanoparticles than CHP.

To investigate the distribution of the hydrophobic domain in the aggregate, the fluorescence quenching technique was applied. The number of hydrophobic units in a polymer micelle was estimated by Poisson quenching kinetics[8]. A hydrophobic fluorescent probe, pyrene (Py), was incorporated into the CHP self-aggregate. The fluorescence of Py is quenched in the presence of cetylpyridinium chloride (CPy^+). In the presence of CHP, the steady-state quenching data for Py by CPy^+ were not consistent with simple Stern-Volmer kinetics, but did fit the following equation:

$$\ln(I_0/I) = [\mathrm{CPy}^+]/[\text{aggregated cholesterol}] \tag{2}$$

where I_0 and I are the fluorescence intensity of Py in the CHP self-aggregate in the absence and presence of CPy^+, respectively. Good straight lines were obtained (r=0.99) and the slope gives 1/[aggregated cholesterol]. In the case of CHP-55-1.7, the mean number of cholesterol moieties per hydrophobic domain, n, can be estimated by eq. 3 to be 4.2 ± 0.5:

$$n = [\text{cholesterol moiety}]_{\mathrm{CHP}}/[\text{aggregated cholesterol}] \tag{3}$$

where $[\text{cholesterol moiety}]_{\mathrm{CHP}}$ is the concentration of the cholesterol moiety in CHP. One CHP-55-1.7 nanoparticle has 58 cholesterol moieties in its self-aggregate. Therefore, there are approximately 14 ± 2 hydrophobic domains per aggregate. A CHP self-aggregate can be considered a nanosize hydrogel in which the microdomain of the associated cholesterol forms non-covalent cross-links (Fig. 2).

The number of crosslinking points in the nanoparticle increased with an increase in the DS of the cholesteryl moiety (Table 1). One hydrophobic domain in the ALP self-aggregate consisted of 12-17 alkyl groups. Thus, there are significantly fewer cross-linking points in the ALP hydrogel nanoparticle (numbers of hydrophobic domain in one nanoparticle: 3-4] than in that of CHP (10-15).

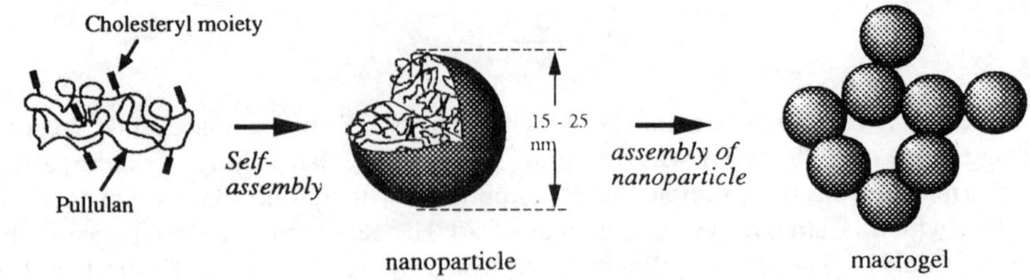

Fig. 2. Schematic representation of self-assembled hydrogel nanoparticle of cholesteryl pullulan

The size, density and stability of hydrogel nanoparticles can be controlled by changing not only the degree of substitution of the hydrophobic group but also by changing their structure or hydrophobicity. This is a new method for preparing hydrogel nanoparticles by the self-organization of hydrophobized polysaccharides.

CHP self-aggregates complex with various hydrophobic substances and provide a chiral binding site for bilirubin[9]. The physicochemical stability of doxorubicin[10] and neocarzinostatin chromophore[11,12] is stabilized by complexation with such nanoparticles.

2.2 Formation of gel

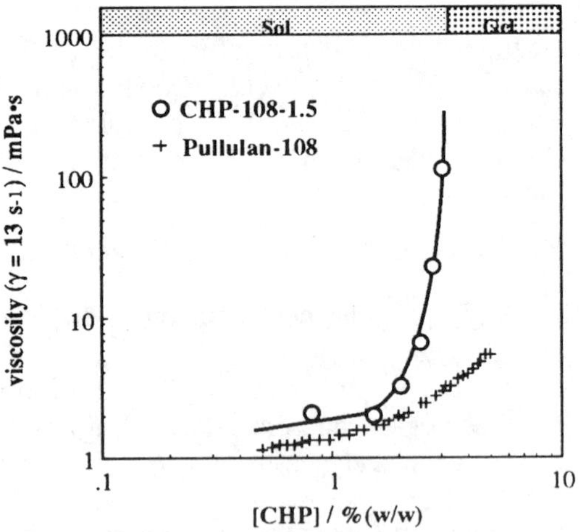

Fig. 3. The viscosity of CHP self-aggregate solution

An aqueous dispersion of nanoparticles of CHP was lyophilized. The resulting white powder was easily dispersible in water at relatively high concentrations. In a semidilute concentration, the viscosity of the CHP self-aggregate solution increased rapidly above approximately 2%(w/w) (Fig. 3). The gelation of CHP was observed above 3.5 %(w/w) by the tilting method[13-15]. A TEM image of a sample of the gel prepared by the freeze-fracturing method shows that the nanoparticles of CHP self-aggregate are interconnected to form a macrogel. Most CHP nanoparticles exist stably even in semidilute solution. On rheological measurements, G" for CHP gel exhibited a maximum at $\omega = 0.1$ rads^{-1} and a slight increase at $\omega = 100$ rads^{-1}, while G' reached a plateau in the high frequency regime[13]. The relaxation of a macrogel is related to the dissociation of associated hydrophobic groups.

2.3 Complexation with proteins and enzyme

Various water-soluble proteins were complexed with CHP self-aggregates[5, 16-20] These complexes show excellent colloidal stability without any precipitation. The concentration of the complexed protein showed a saturation phenomenon as a function of the initial concentration of the protein added. One nanoparticle of CHP-55-1.7 complexes with one protein for bovine serum albumin (BSA , Mw=67000), one for α-chymotrypsin dimer (Mw=49000), three for interferon a (Mw=18750), four for cytochrome c (Mw=12480) and 10 for insulin (Mw=5735). The maximum amount of protein bound to CHP nanoparticles depends on the molecular weight (or size) of the protein. CHP nanoparticles seem to have a limited capacity for proteins of a certain size. The physicochemical and thermal stability of proteins is drastically increased by complexation.

The spontaneous dissociation of the protein from the complex scarcely occurs even after a week at 25 °C. However, complexed proteins are released by exchange with another protein which is thermodynamically more easily complexed with CHP. For example, insulin (Ins) was released from the complex with the addition of BSA[19]. The release of Ins depended on the BSA concentration. α-chymotrypsin (Chy) was released from the complex with the addition of BSA[16]. The released Chy had almost the same enzymatic activity as that of free Chy before complexation, although the enzymatic activity of the complexed Chy was decreased approximately 100-fold.

Complexed protein is also released with the addition of β-cyclodextrin and the accompanying dissociation of the nanoparticles[20].

Hydrogel nanoparticles of CHP self-aggregates can be considered a host for a macromolecular guest such as protein. This is an example of well-controlled macromolecular association between two water-soluble macromolecules.

2.4 Artificial molecular chaperone: Thermal stabilization of enzymes by association with macromolecules

Molecular chaperones selectively trap heat-denatured proteins and intermediate in refolding to prevent irreversible aggregation[21]. The host chaperone releases the complexed protein upon cooling, and this is accompanied by a change in the conformation of the complex with the aid of ATP and a co-chaperone. The denatured protein refolds to a native form. This is an interesting example of host-guest chemistry in macromolecular systems. We can simulate the two-step mechanism of a molecular chaperone (catching and releasing of a protein) in heat-induced denaturation using CHP nanoparticles and β-cyclodextrin[20, 22]. The irreversible aggregation of an enzyme such as carbonic anhydrase B (CAB) upon heating was completely prevented by the selective and quantitative complexation of the heat-denatured enzyme with nanoparticles of self-aggregated CHP. The complexed CAB was released by the addition of β-cyclodextrin. The denatured CAB refolded to a native form, and almost 100% of the activity was recovered. The thermal stability of CAB was drastically improved by the catching and releasing of CAB with CHP nanoparticles. This is a novel method for the thermal stabilization of enzymes using an artificial host based on a mechanism similar to that of a molecular chaperone (Fig. 4).

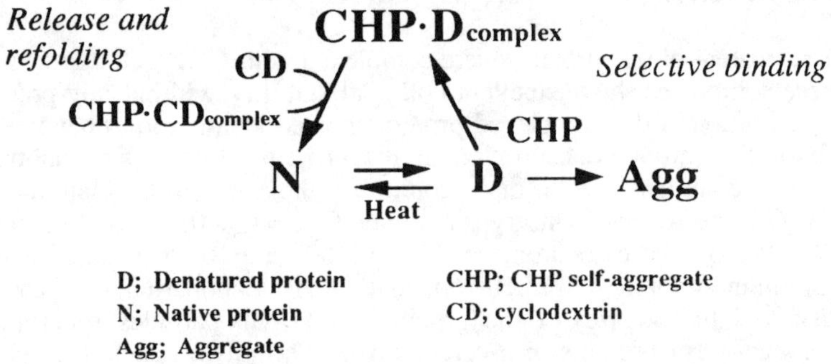

Fig. 4. Schematic representation of thermal stabilization of enzyme by CHP

3. Self-assembly of hydrophobized poly(amino acids)

The partial modification of polysaccharides by a very hydrophobic group such as cholesterol leads to the formation of stable monodispersive hydrogel nanoparticles in water. To investigate the generality of this method, we selected a polyelectrolyte, poly(L-lysine) (PLL), as a parent polymer instead of nonionic polysaccharides. We focused on the self-aggregation of cholesteryl-bearing poly(L-lysine) (CHPLL) in water and the effect of association with hydrophobes on the secondary structure of CHPLL[23].

CHPLL was synthesized by the condensation of cholesteryl N-(6-isocyanatehexyl) carbamate with PLL (Fig. 5). Sonicated samples of all of the CHPLLs, which were substituted by 0.8, 2.2, 3.4 and 5.4 cholesterol groups per 100 lysine units, gave a single peak after sufficient ultrasonication by high-performance liquid chromatography. Spherical nanoparticles of relatively uniform size were observed by electron microscopy. The particle sizes (R_G) and aggregation numbers (N), which were determined by static light scattering, increased with an increase in the DS of cholesterol of CHPLL (R_G = 16nm - 22nm, N = 1.3 - 4.2).

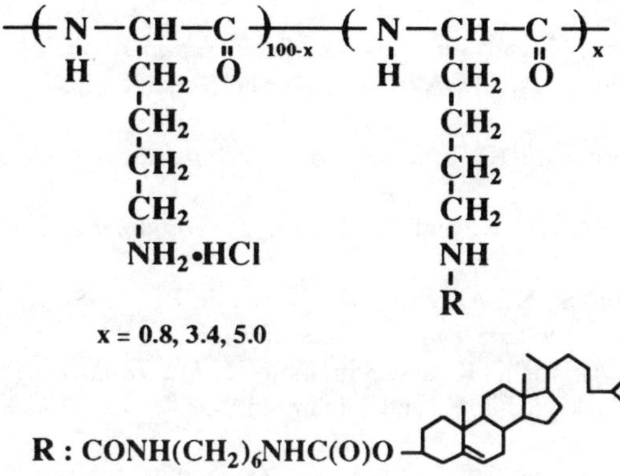

Fig. 5. Chemical structure of cholesteryl-bearing poly(L-lysine) (CHPLL)

PLL forms unique conformations in solution, such as α-helix and β-sheet structures, depending on the pH[24]. It is easy to detect conformational changes in PLL by circular dichroism (CD). CHPLLs with a higher DS of cholesterol groups tended to form an α-helical structure at lower pH by CD spectroscopy. The partial modification of PLL by the strongly hydrophobic cholesteryl group led to the formation of hydrogel nanoparticles by their self-assembly and induced an α-helical structure for PLL at lower pH.

Acknowledgment: The authors gratefully acknowledge financial support from a Grant-in-Aid for Scientific Research from the Ministry of Education, Science, and Culture and Terumo Life Science Foundation.

References

1. Shalaby, S. W., McCormick, C. L. and Butler, G. B. Eds. In *Water-Soluble Polymers*; American Chemical Society: Washington, DC, 1991.
2. Dubin, P., Bock, J., Davis, R., Schulz, D.N. and Thies, C., Eds. In *Macromolecular Complexes in Chemistry and Biology*; Springer-Verlag: Berlin, 1994.
3. Morishima, Y. *Trends Polym. Sci.* 1994, **2**, 31.
4. Akiyoshi, K., Deguchi, S., Moriguchi, N., Yamaguchi, S. and Sunamoto, J. *Macromolecules* 1993, **26**, 3062.
5. Akiyoshi, K. and Sunamoto, J. Supramolecular Science, 1996, **3**, 157.
6. Akiyoshi, K., Deguchi, S., Tajima, T., Nishikawa, T. and Sunamoto, J. *Macromolecules* 1997, **30**, 857.
7. Kato, T., Katsuki, T. and Takahashi, A. *Macromolecules* 1984, **17**, 1726.
8. Zana, R. In *Surfactant Solutions: New Methods of Investigation*; Zana, R., Ed.; Marcel Dekker: New York, 1987, **22**, 241

9. Akiyoshi, K., Deguchi, S., Tajima, H., Nishikawa, T. and Sunamoto, J. *Proc.Japan Acad.*, 1995, **71**, Ser. B, 15
10. Akiyoshi, K., Taniguchi, I., Fukui, H. and Sunamoto, J. *Eur. J. Pharma. Biopharm.*, 1996, **42**, 286.
11. Kato, Y., Sugiura, Y. and Sunamoto, J. *Proc. Japan Acad.*, 1998, **74**, Ser. B, 116.
12. Taniguchi, I., Fujiwara, M., Akiyoshi, K. and Sunamoto, *J. Bull. Chem. Soc. Jpn.*, 1998, **71**, 2681.
13. Deguchi, S., Akiyoshi, K., Lindman, B. and Sunamoto, J. *Macromolecular Symp.*, 1996, **109**, 1.
14. Akiyoshi, K., Kuroda, K. and Sunamoto, J. *Kobunnshi Ronbunshyu*, 1998, **55**, 781.
15. Deguchi, S., Kuroda, K., Akiyoshi, K., Lindman, B. and Sunamoto, J. *Colloids and Surfaces A: Physicochem. Eng. Aspects*, in press.
16. Nishikawa, T., Akiyoshi, K. and Sunamoto, J. *Macromolecules*, 1994, **27**, 7654.
17. Nishikawa, T., Akiyoshi, K. and Sunamoto, J. *J. Am. Chem. Soc.*, 1996, **118**, 6110.
18. Akiyoshi, K., Nishikawa, T., Mitsui, Y., Miyata, T., Kodama, M. and Sunamoto, J. *Colloids and surfaces A: Physicochem. Eng. Aspects*, 1996, **112**, 91.
19. Akiyoshi, K., Kobayashi, S., Shichibe, S., Mix, D., Baudys, M., Kim, S. W. and Sunamoto, J. *J. Contr. Release*, 1998 **54**, 313.
20. Akiyoshi, K., Sasaki, Y., Kuroda, K. and Sunamoto, J. *Chem. Lett*, 1998, 93.
21. Mayhew, M., da Silva, A. C. R., Martin, J., E-bromage, H., Tempst, P. and Hartl, U. *Nature*, 1996, **379**, 420.
22. Akiyoshi, K., Sasaki, Y. and Sunamoto, J. Bioconjugate Chem. submitted publication.
23. Ueminami, A. Akiyoshi, K., and Sunamoto, J. *Polym. Prep. Jpn.*1995, **44**, 2617.
24. Greenfield, N. and Fasman, G. D. *Biochemistry*, 1969, **8**, 4108.

Macro to Microdynamics of Threadlike Micelles

Toshiyuki Shikata

Department of Macromolecular Science, Osaka University
Toyonaka, Osaka 560-0043, Japan
shikata@chem.sci.osaka-u.ac.jp

Abstract. The dynamics of the entangling threadlike micelles formed with surfactants and additives in aqueous solution can be classified into three regimes. The first dynamics is the slowest one related to the entanglement release mechanism of threadlike micellar networks. The second dynamics is governed by bending motions of the threadlike micelle between entanglement points. These are *macrodynamics* of the threadlike micelle, which can be investigated by rheological techniques. The last dynamics of the threadlike micelle is *microdynamics*, which is in time scale on the order of 10^{-9} s and is detectable by use of fluorescence probing techniques. Fast molecular motions for surfactants and additive molecules in the micelle are the origin of this dynamics. Rotational and translational motions for surfactants and additives in the threadlike micelle are twice slower than those in the spherical micelle.

1. INTRODUCTION

1.1. Macrodynamics of Threadlike Micelles

It is well known that some cationic surfactants from enormously long threadlike micelles in aqueous solution with additives.[1-3] The long threadlike micelle makes concentrated entanglement networks and shows pronounced viscoelastic behavior as polymer molecules do in semi-dilute to concentrated conditions. From the viewpoint of viscoelastic features, the threadlike micelle behaves as a flexible long thread with many entangling points possessing a finite lifetime.[3, 4] However, if one pays attention to the viscoelastic behavior of the entangling threadlike micellar system very carefully, especially to the slowest relaxation mechanism related to the entanglement release, one finds out obvious difference from that of the concentrated polymer system. In the concentrated polymer system, the slowest relaxation mode shows a box type relaxation spectrum which means the broad distribution of relaxation modes.[5] On the contrary, the slowest relaxation mode of the threadlike micellar system in a time scale on the order of 1 s show a Maxwell element like very sharp relaxation spectrum, which means it possesses single relaxation time in the slowest mode.[3] This

Y. Morishima · T. Norisuye · K. Tashiro (Eds.)
Molecular Interactions and Time-Space Organization in Macromolecular Systems
Proceedings of the OUMS '98, Osaka, Japan, 3-6 June, 1998

difference in the slowest relaxation mechanism reflects essential difference in structural features of these threadlike matters. In the case of the polymer chain molecule, chain units are tightly bound each other by the covalent chemical bonding. On the other hand, the threadlike micelle is formed with not so strong intermolecular interaction between surfactant molecules called the *hydrophobic interaction*. Therefore, it is possible that two threadlike micelles cross and pass through each other at an entanglement point between themselves like phantoms, because the micelle forming hydrophobic interaction is much weaker than the covalent chemical bonding.

We have proposed a model[4] called *a phantom crossing model* to understand the unique viscoelasticity of the threadlike micellar system. In the model we assume that every entanglement point has a lifetime equal to the mechanical relaxation time, and two threadlike micelles pass through each other at an entanglement point when the lifetime is over. Other groups investigating physical properties of the threadlike micelle also have proposed different models based on an idea in which the scission and the reformation of the threadlike micelle occur at every portion of the micelle.[2]

When one looks at dynamics of the threadlike micelle between entanglement points in a time scale from 10^{-1} to 10^{-3} s much shorter than the slowest relaxation time, one can not find out any differences in the viscoelastic behavior of the concentrated polymer system and of the entangling threadlike micellar system.[6] The reason why no differences in the viscoelasticity of the two systems might be that both the polymer chain and the threadlike micelle behave as only flexible threadlike matter which can bend rather freely.

These dynamics of the threadlike micelle related to the entanglement release and bending motion of the micelle are *macrodynamics*, which are detected by rheological techniques.

1.2. Microdynamics of Threadlike Micelles

The last part of consideration about dynamics of the threadlike micelle is *microdynamics*.[7] In the threadlike micelle, surfactants and additives should have considerably fast molecular motions. They always alter their positions and rotate quickly to reduce the memory of position and orientation in the micelle, since the micelle forming origin is the intermolecular interaction not the chemical bonding . In this paper, the microdynamics of the threadlike micelle will be discussed mainly. By choosing sophisticated molecules which can probe the molecular motions of the micelle forming substances, the microdynamics of the threadlike micelle can be revealed. A fluorescence anisotropy relaxation measurement was employed to investigate rotational relaxation times of the fluorescent probe molecules working as substitutes for micelle forming surfactant and additive molecules.

A combination between cetyltrimethylammonium bromide (CTAB) and sodium salicylate (NaSal)[3] has been frequently employed as micelle forming substances so far for investigation of several physicochemical features in threadlike micellar system. However, the fluorescence of NaSal is too strong to examine the fluorescence behavior of molecular probes incorporated in the threadlike micelle.[7] Thus, in the investigation of the microdynamics of the threadlike micelle with the fluorescence techniques, sodium *p*-toluenesulfonate (Na*p*TS), which is very weakly fluorescent, was used as an additive to enhance the threadlike micelle formation of CTAB.

Threadlike micelles formed in an aqueous solution of CTAB and Na*p*TS (CTAB:Na*p*TS/W) are known to exhibit similar rheological behavior to that observed in the aqueous CTAB and NaSal system.[8] Thus, the microdynamics of threadlike micelles formed in the CTAB and Na*p*TS system might be essentially the same as that in the CTAB and NaSal system.

2. FLUORESCENCE ANISOTROPY RELAXATION

2.1. Materials and Methods

Fluorescent molecules, sodium 2-hydroxy-3-naphthoate (NaHN), 1-[4-(trimethylamino)phenyl]-6-phenylhexa-1,3,5-triene bromide (TMA-DPH) were commercially available. Another fluorescent probe molecule, cetylacridiniumorange bromide (CAOB), was synthesized according to a method proposed by Miethke and Zanker[9], and was recrystallized in methanol.

The concentration of CTAB (C_D) was kept at 3, 10 and 50 mM, while the concentration of Na*p*TS (C_S) was altered from 0 to 300 mM. The concentrations of NaHN was about 10 μM and were much lower than the C_D value. Moreover, the concentrations of TMA-DPH and CAOB were lower than 1μM.

Steady state fluorescence anisotropy (r) of the fluorescent molecular probes incorporated in the micelle was determined at 25 ºC with a conventional fluorescence photometer equipped with a polarizer in front of the excitation window and an analyzer in front of the emission one. Excitation wave lengths for the probes, NaHN, CAOB and TMA-DPH, were 365, 498 and 360 nm, respectively. Furthermore, emission wave lengths for these probes were 512, 512 and 433 nm, respectively. The value of r was estimated with eq 1 below from fluorescence intensities, I_{vv} and I_{vh} : The subscripts (v; vertical and h; horizontal) of the intensities mean the direction of the polarizer and of analyzer, respectively. The factor of g means the correction factor for the polarizing character of the fluorescence photometer used and can be estimated by $g = I_{vh}/I_{hh}$.

$$r = \frac{I_{vv} - g\, I_{vh}}{I_{vv} + 2g\, I_{vh}} \tag{1}$$

Fluorescence lifetimes (τ_{life}) of the molecular probes in the micelle were measured by use of a conventional time correlation single photon counting apparatus equipped with a high pressure hydrogen flash lamp. The half width of the flash pulse was ca. 2 ns. A deconvolution program was operated to calculate the τ_{life} values correctly.

Rotational relaxation times (τ_ϕ) of the fluorescent molecular probes were evaluated with eq 2 through the values of r, τ_{life} and instantaneous fluorescence anisotropy (r_0), which was determined in an extraordinary viscous medium like glycerin at low temperature.

$$\frac{1}{r} = \frac{1}{r_0}\left(1 + \frac{3\tau_{life}}{\tau_\phi}\right) \qquad (2)$$

A spherical micelle can be regarded as a rigid sphere in a time scale longer than the microdynamics of the micelle. If one assumes a radius of the spherical micelle of CTAB in aqueous solution to be 2.5 nm[10], the rotational relaxation time for the spherical micelle can be evaluated as ca. 50 ns with the Stokes-Einstein relationship. Moreover, the time constant for over all rotation of the spherical micelles or bending motion of the threadlike micelle should be equal to or longer than this value. By the way, the values of τ_ϕ for all the fluorescent molecular probes in this study *via* eq 2 were shorter than 5 ns.

Here, one can consider a simple case that a probe molecule is incorporated in a spherical micelle and has the rotational relaxation time (τ_ϕ^m) for the microdynamics, and the micelle has the rotational relaxation time (τ_ϕ^r) for the over all rotation. In this case, the value of τ_ϕ can be expressed in a manner of $1/\tau_\phi = 1/\tau_\phi^m + 1/\tau_\phi^r$. From this consideration, the τ_ϕ value in the case of this study essentially corresponds to τ_ϕ^m, because the τ_ϕ^r value; ca. 50 ns, is much longer than τ_ϕ^m.

2.2. Fluorescent Molecular Probe for *p*TS$^-$

Fluorescent probe molecules, NaHN, is incorporated in the micelle in the form of dissociated anions, HN$^-$. This is confirmed by fluorescence lifetimes, τ_{life}, measurement. The τ_{life} value for NaHN in the pure water was 1.5 ns and 5~6.5 ns in the micelle. From the NMR investigation[8] of a CTAB and Na*p*TS system in D_2O, it was revealed that in the threadlike micelle Br$^-$ of CTAB was replaced by dissociated *p*TS$^-$ from Na*p*TS. The composition between CTA$^+$ and *p*ST$^-$ in the threadlike micelle was determined to be unity. Thus, the ratio of C_S /C_D can be a good parameter indicating the shape of micelles. A region with the C_S /C_D value from 0 to unity is occupied by spherical to short rodlike micelles. On the other hand, a region with C_S /C_D >1 is occupied by long threadlike micelles grown enough.

Because the chemical structures of NaHN is not so different from that of Na*p*TS, it might play a role of a substitute for Na*p*TS in the micelle. NaHN has two hydrophilic groups such as OH and COO$^-$, so that it can be incorporated in the micellar interior sticking out these two hydrophilic groups toward the bulk aqueous phase as schematically shown in Figure 1. The direction of the transition moment of NaHN is aligned as seen in the same figure schematically. The value of r_0 for NaHN was determined to be 0.34 in glycerin at -10 ºC, and was not so different from the limiting value of 0.4 at the condition for which the transition moment and the axis of fluorescence emission was parallel. Therefore, the rotational motion around an axis shown in Figure 1 may be the most effective motion to relax the fluorescence anisotropy.

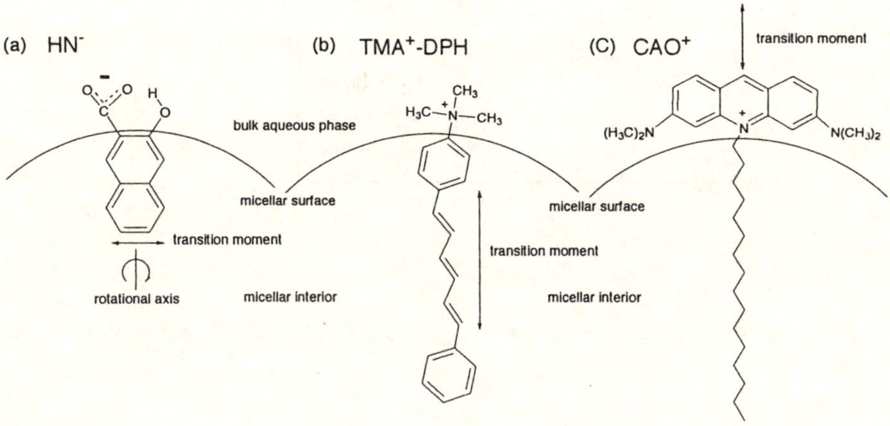

Fig. 1. Schematic representation of molecular structures, directions of transition moments and axis of rotation for HN^-, TMA^+-DPH and CAO^+ ions incorporated in the micelle.

The relationship between τ_ϕ for NaHN incorporated in the micelle and a ratio of C_S /C_D is plotted in Figure 2. Stepwise increase in τ_ϕ data around C_S /C_D =1 for several C_D in Figure 2 implies a change in the rates of molecular motions of the HN^- anion in the micelle caused by a change in the shape of micelle. Because the τ_ϕ value for HN^- incorporated in the micelle implies the rotational relaxation time around the axis shown in Figure 1 as pointed out above, the data in Figure 2 mean that the rotational rate of HN^- in the threadlike micelle is about a half of that in the spherical one. This fact reflects difference in the inner viscosities of the two kinds of micelles, since the rate of the rotational Brownian motion is inversely proportional to the viscosity of a medium surrounding the probe molecule. Here, we conclude that the inner viscosity of the threadlike micelle is twice higher than that of the spherical one.

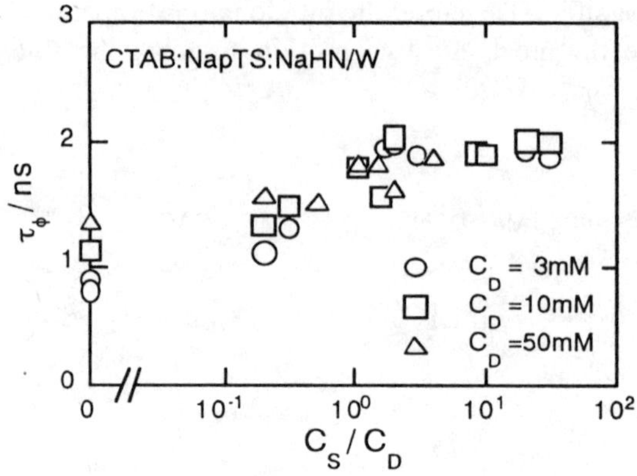

Fig. 2. The dependence of τ_ϕ on the C_S/C_D ratio for HN^- in the CTAB:*Nap*TS/W system at 25 °C

2.3. Fluorescent Molecular Probes for CTA^+

TMA-DPH and CAOB have chemical structures not so different from that of CTAB. TMA-DPH has the fluorescence lifetime, τ_{life}, shorter than 0.3 ns in water, whereas its τ_{life} becomes 1~1.5 ns in the micellar interior. On the other hand, CAOB is insoluble in water, however, it can be well incorporated in the micelle formed with CTAB and shows the τ_{life} value of 1~1.5 ns. Therefore, both TMA-DPH and CAOB do work well as substitute fluorescent probe molecules in the micelle in dissociated forms, TMA^+-DPH and CAO^+.

Because the direction of the transition moment of TMA-DPH is aligned in the long molecular axis connecting two phenyl rings as shown schematically in Figure 1 and the value of r_0 is 0.39, the effective molecular motion in the micelle for the fluorescence anisotropy relaxation is the translational motion along the curvilinear micellar surface and/or the wobbling motion. On the other hand, the direction of the transition moment of CAO^+ is parallel to an axis connecting nitrogen and 9-carbon atom in an acridinium ring as also seen in Figure 1. Thus, the translational motion along the micellar surface and/or the wobbling one should be also effective to relax the fluorescence anisotropy for CAO^+ in the micellar interior.

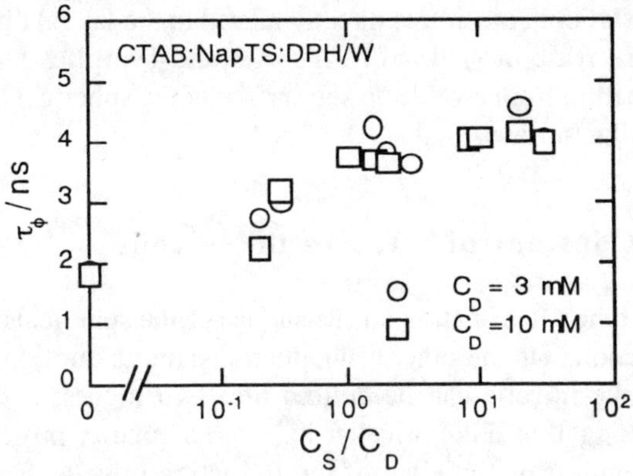

Fig. 3. The dependence of τ_ϕ on the C_S/C_D ratio for TMA^+-DPH in the CTAB:NapTS/W system at 25 °C

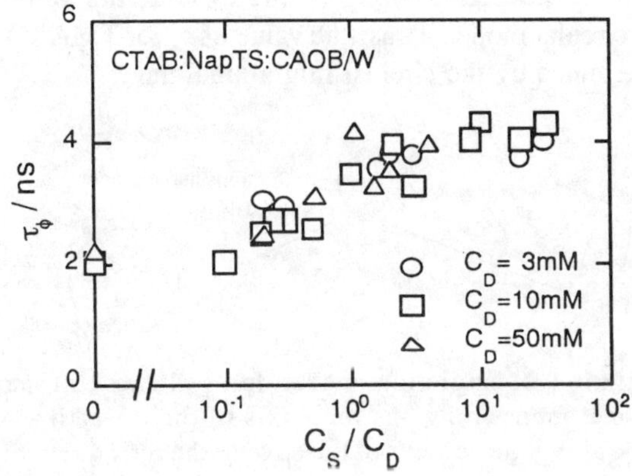

Fig. 4. The dependence of τ_ϕ on the C_S/C_D ratio for CAO^+ in the CTAB:NapTS/W system at 25 °C.

The τ_ϕ values for TMA^+-DPH and CAO^+ are plotted as functions of the C_S/C_D ratio in Figures 3 and 4. The C_S/C_D ratio dependence of the data in these figures roughly resemble that in Figure 2. This means that rates of molecular motions for both TMA^+-DPH and CAO^+ alter with the change of shape of the micelle due to the C_S/C_D ratio as well as HN^-. Therefore, we conclude that in the threadlike micelle the translational motion along the micellar surface and/or the wobbling motion of the CTA^+ cation is twice slower than in the spherical micelle.

Because the rate of the translational motion is directly related to the lateral diffusion constant (D_{lat}), the stepwise increase in τ_ϕ data of Figures 3 and 4 implies stepwise decrease in D_{lat} for the CTA^+ cation from a value on the surface of the spherical micelle to that on the threadlike micellar surface.

2.4. Lateral Diffusion Constant of CTA^+ in the micelle

The translational motion of the CTA^+ cation on the surface of the spherical micelle always occurs in random direction. On the other hand, the translational motion on the rodlike surface of the threadlike micelle can be divided into two modes; a straight migrational mode along the long threadlike micellar axis and a circular migrational mode along the curvilinear contour of the slice of the threadlike micelle as shown schematically in Figure 5. When the substitute fluorescence molecules such as TMA-DPH and CAOB have the transition moment parallel to the long molecular axis, the circular migrational mode must be effective to relax the fluorescence anisotropy. However, the straight mode is not effective to relax it, because the direction of the transition moment does not alter with time. If the probe molecules use only the straight migrational mode , they need much longer time to relax the fluorescence anisotropy than they use the circular mode. Thus, the value of τ_ϕ for TMA^+-DPH and CAO^+ can be essentially determined by the circular migrational mode.

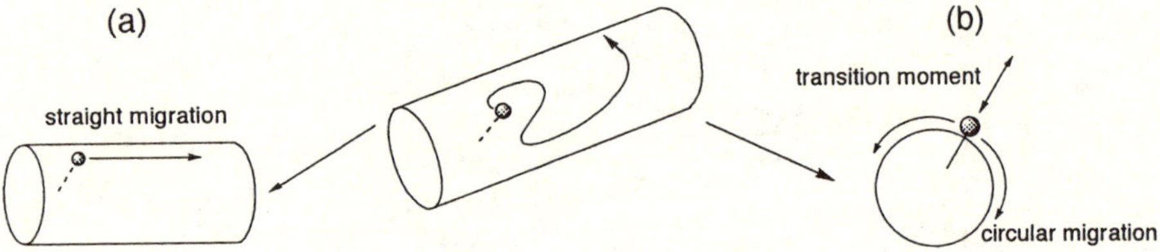

Fig. 5. Schematic representation of migration modes for a fluorescent molecular probe, (a); a straight migrational mode along the long axis of the threadlike micelle, (b); a circular migrational mode along the contour of the slice of the threadlike micelle.

If one neglects a numerical front factor for simplicity, D_{lat} of the fluorescence probes or the CTA^+ cation in the threadlike micelle can be roughly estimated by eq 3 with the radius (a) of the threadlike micelle.

$$D_{lat} \approx \frac{a^2}{\tau_\phi} \qquad (3)$$

By use of eq 3 assuming $a = 2.5$ nm, the D_{lat} values for CTA^+ in the spherical and threadlike micelle can be estimated to be 2×10^{-5} and 1×10^{-5} cm^2s^{-1}, respectively, from Figures 3 and 4. These values are larger by a factor of five to fifty than the reported D_{lat} values for surfactants in micellar systems estimated with different techniques by other groups so far.[11,12] This discrepancy will be explained by some factors which has not been considered yet in this study and also in the cited work [10,11]. Presumably two factors below will be necessary for further discussion on the D_{lat} value. The first factor is the neglected numerical front factor, and the other is the contribution of the quick wobbling motion of the probe molecules to the fluorescence anisotropy relaxation.

If a time resolved fluorescence anisotropy relaxation experiment with a very short width excitation pulse is carried out in this system, the contribution of quick wobbling motion to the fluorescence anisotropy relaxation will be separated from the total fluorescence anisotropy decay curves and the pure contribution of the translational migration mode of the probe molecules will be detected exactly.[13]

3. CONCLUSIONS

The dynamics of the entangling threadlike micellar system can be classified into three regimes. The first one is the slowest dynamics observed in the time scale of 1 s. The mechanism for entanglement release of threadlike micelles is included in this regime. The second dynamics of the threadlike micellar system is bending motions of the threadlike micelle between entanglement points. The time scale of this dynamics is on the order of 10^{-1} to 10^{-3} s. These macroscopic dynamics are investigated with rheological techniques. The last dynamics of the threadlike micelle is microdynamics, which has a time scale on the order of 10^{-9} s and is governed by molecular motions for surfactants and additive molecules in the micelle. Rotational and translational motions for both surfactants and additives in the threadlike micelle are twice slower than those in the spherical micelle

References

1) Glavsholts, S., *J. Colloid Interface Sci.*, **57**, 575 (1976).
2) Hoffmann, H. and H. Rehage, *Mol. Phys.*, **5**, 1225 (1989).
3) Shikata, T., H. Hirata and T. Kotaka, *Langmuir*, **3**, 1081 (1987).
4) Shikata, T., H. Hirata and T. Kotaka, *Langmuir*, **4**, 354 (1988).
5) Ferry, J. D., *Viscoelastic Properties of Polymers, 3rd ed.*, Wiley, New York, 1980.
6) Shikata, T. in preparation.
7) Shikata, T. and Y. Morishima, *Langmuir*, **13**, 5229 (1997).
8) Shikata, T., unpublished data.
9) Miethke, E. and V. Z. Zanker, *Phys. Chem.* (Frankfurt am Main), **18**, 375 (1958).
10) Shikata, T. and S. Imai, *Langmuir*, in press 1998.
11) Söderman, 0., U. Henriksson and U. Olsson, *J. Phys. Chem.*, **91**, 116 (1987).
12) Alsins, J. and M. Almgren, *J. Phys. Chem.*, **94**, 3062 (1990).
13) Kinoshita, S., I. Tanaka, T. Kushida, Y. Kinoshita and S. Kimura, *J. Phys. Soc. Jpn.* **51**, 598 (1982).

Self-Organized Nanostructures with a Single Polymer Chain

Kenichi Yoshikawa

Department of Physics, Graduate School of Science, Kyoto University, Kyoto 606-8502, Japan
E-mail: yoshikaw@scphys.kyoto-u.ac.jp

Abstract. The present article describes the new finding that various kinds of characteristic nanostructures are created spontaneously from the collapsing transition in single polymer chains. From theoretical study with Monte-Carlo simulation on single chains, it is predicted that toroid, rod, and spherical structures are generated as the most stable conformations under thermal equilibrium, depending on the stiffness, temperature (or effective interaction between the segments in a chain), and chain length. From the actual experiments on long DNA chains with the order of several tens of micrometers, the formation of these nonostructures have been confirmed, together with the formation of intra-chain segregated structure. It is also indicated that kinetic effect, such as nucleation-growth and spinodal decomposition, plays an important role in the determination of the final structures with the collapsing transition.

Introduction

In good solvent a polymer chain takes elongated coil conformation, whereas in poor solvent collapsed globule conformation is preferred[1, 2, 3]. Between the two extremes of the conformation, there lays the region of the transition. The transition between the elongated state and the collapsed state has usually been called as coil-globule transition. Almost all of the past experimental studies on the coil-globule transition have indicated that the transition is always continuous, although in stiff polymers, including DNA, the transition curve becomes very sharp[4]. Such a continuos character in the transition has been discussion in relation to various possible factors; the finite size effect of the polymer chain, kinetic effect, order of the transition, etc. As for the order of the transition, deGennes[2] suggested that this transition will approach a criticality, as the number of segments N of the molecular chain length is increased to infinity as in the critical behavior in second-order phase transition. This discussion may imply the expectation that no discreteness would be found in the order parameter such as the segment density with respect to control parameters such as temperature and solvent quality.

Y. Morishima · T. Norisuye · K. Tashiro (Eds.)
Molecular Interactions and Time-Space Organization in Macromolecular Systems
Proceedings of the OUMS '98, Osaka, Japan, 3-6 June, 1998

Recently, it has been revealed that isolated single long DNA chains exhibit a marked discrete transition with the difference of the density of the order of 10^3-10^4 between the elongated and collapsed state, through the experimental observation on the conformation of single DNA chains by use of fluorescence microscopy[5, 6, 7, 8]. Contrary to this large discrete change in the transition on the level of individual chains, the ensemble average of DNAs shows a steep but continuous transition. In usual aqueous solution, a DNA chain has a double-stranded structure and its persistence length is 50-60nm. It is noted that this persistence length is 25-30 times larger than its diameter, 2nm. Natural DNA chains have lengths that are several orders of magnitude larger than the persistence length; 0.1-1nm in bacterial phage, 1-10nm in bacteria and 1-10cm in eukaryote. An aqueous solution is a good solvent for highly charged DNA chains; long natural DNAs behave as semiflexible polymer chains and exhibit an elongated coil state in aqueous environment. When the quality of the aqueous solution is changed worse by adding a low-polar solvent[9] (or multivalent cations[7, 11], or flexible hydrophilic polymer[11, 12], or cationic surfactant[13,14]), individual long DNA chains collapse, one by one, into collapsed compact state, reflecting the marked discrete character of the transition.

Currently, the term of the so-called globule seems to include the meaning of spherical shape[3]. In the present article, we will show that there exist various kinds of characteristic morphologies in the collapsed products from single chains, both from theoretical calculation and experimental observation on DNAs. In liquid like collapsed state, the morphology in the free energy minimum is spherical globule, in which structure the surface energy is minimized. Whereas, as will be described below, exotic nanostructures will be generated spontaneously from stiff polymer chains.

Phase Diagram of Single Polymer

In Fig. 1 is shown the schematic representation of the phase diagram for single polymer chain as has been obtained through Monte Carlo calculation with multicanonical algolism[15]. With this methodology, one can obtain the stable structures under thermal equilibrium. As for this theoretical approach, the reader can find the detail in the published articles[15, 16, 17]. The longitudinal axis in the figure indicates the change of the solvent quality. In stead of the change in solvent quality, one can regard the longitudinal axis as the change in temperature; the upper region corresponds to higher temperatures. (As will be discussed in the next section, polyelectrolyte chain can perform the transition from globule to coil with the increase of temperatures, as in the case of the transition of DNA induced by polyamines. In such case, one can re-read the diagram with the reflection of the temperature axis.)

Thus, the coil state existing in upper region of the diagram corresponds to a kind of "gas" phase. In flexible polymer, with the decrease of the temperature the gas-like coil state collapses into liquid-like spherical state. When the temperature decreases much more, the liquid state changes into the solid state. On the other hand, for stiff polymer the coil state changes into solid-like toroidal structure with the

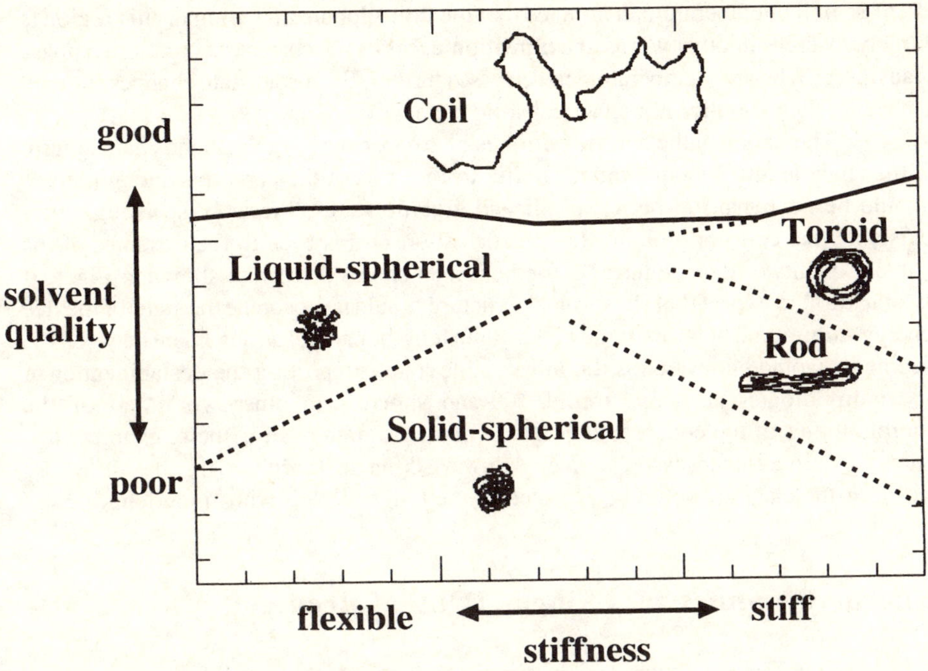

Fig. 1. Schematic representation of the phase diagram in the morphological change of single polymer chain with the parameters of chain stiffness and solvent quality. (Partly modified from Ref. 15)

decrease of the temperature. It is noted that the coil-globule transition in this region is largely discrete, in other words, the transition exhibits the character as first-order phase transition. When the temperature is decreased further, the toroid state changes into rod state and, then, finally into spherical solid state.

The nature of the coil transition is also dependent largely on the chain length. If the chain length becomes much shorter to the order of the persistence length, there should be no transition between coil and globule states. Precisely speaking, there exists neither coil nor globule state in such short chains even in the extremes of the solvent quality or temperature. As for the longer chains than those shown in Fig.1, it is of natural to expect that the toroidal structure would not become the state of the free energy minimum. In the extreme of the chain length, parallel arrangement between the segments would be favored as the most stable state, suggesting the destabilization of the morphologies such as toroid, rod and sphere. In other words, all of the morphologies of the collapsed state given in the diagram of Fig.1 belongs to the free energy minima in mesoscopic state. Anyway, different morphologies besides those given in the diagram would be expected in the longer chains with finite length.

Thermodynamics in a Single DNA Molecule

The discrete character of the transition between elongated coil and collapsed globule is clearly confirmed with the single chain observation of long DNA molecules with use of fluorescence microscopy. Fig.2 shows fluorescence microscopic images of T4DNA (166kbp and 57um in contour length) stained with a fluorescent, 4',6-diamidino-2-phenylindole (DAPI), in the presence of 1.2mM spermidine at different temperatures, together with corresponding quasi-three-dimensional representations where the vertical scale represents the intensity of the fluorescence[18]. It is noted that with an increase of the temperature, the relative population of the collapsed DNA chains increases.

Fig.3 shows the distribution of the long-axis length at different temperatures in the fixed concentration of spermidine (1.2mM). In this histogram, the coil and globule states have been discriminated in a definite manner based on the observation of time-dependent trace in the fluorescence images, i.e., intrachain and translational Brownian motion of individual chains.

From the experimental results as are shown in Figs.2 and 3, change in entropy in the first-order phase transition of single T4 DNA is evaluated as: $\Delta S = 32$ k at $T_c = 291$ K, where k is the Boltzman constant ($k = 1.32\ 10^{-23}$ J/K). The phase transition temperature, T_c, is deduced from the temperature with the 50 :50 distribution between coil and globule. The enthalpy change per single chain is, thus, calculated as $\Delta H = 1.2\ 10^{-19}$ J at T_c. The physico-chemical consideration on these thermodynamic parameters will be described in a forthcoming article. At the present, we would like to mention that the change in the translational entropy of the counter ions plays the most significant role in the coil-globule transition induced by polycations such as spermidine[7]. As the entropy gain due to the ion exchange between monovalent ions,

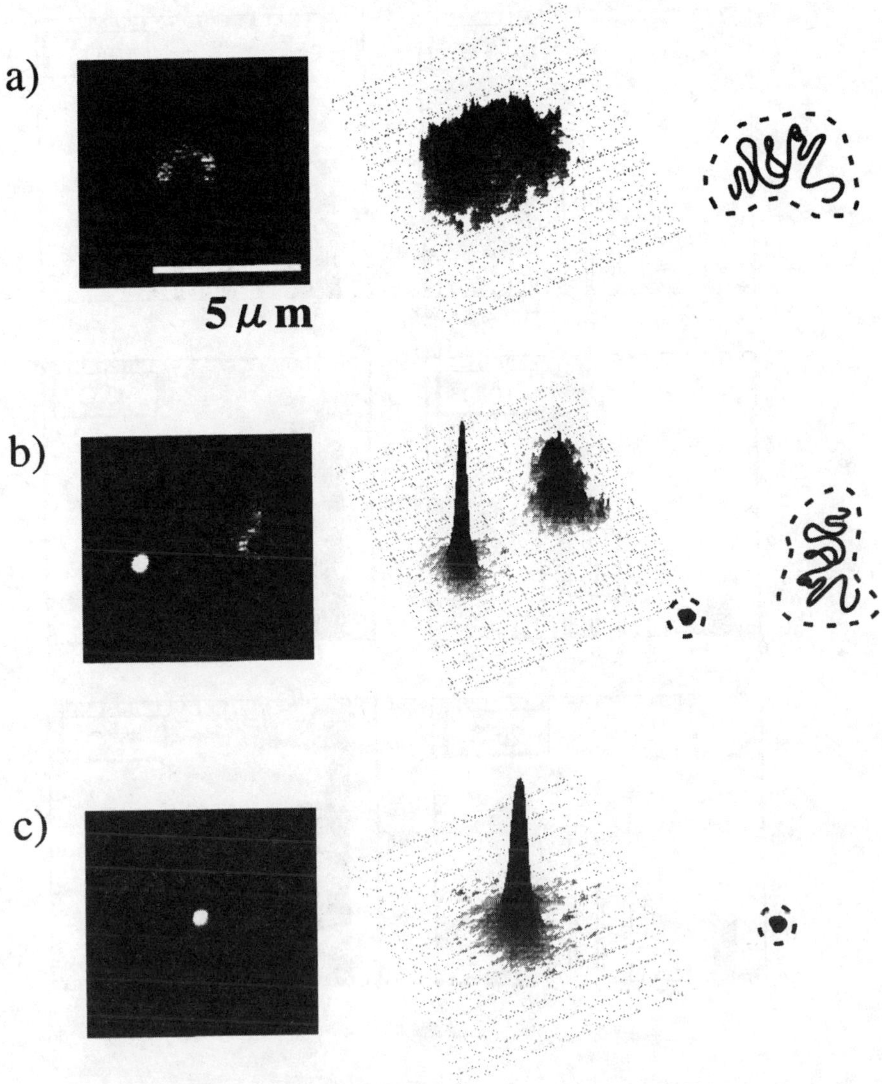

Fig. 2. Left: Fluorescence images of T4 DNA molecules at different temperatures. Middle: Corresponding quasi-three-dimensional representations of the fluorescence images, where the vertical scale indicates the fluorescence intensity. Right: Schematic representation of the DNA conformations, where the broken line indicates the blurring effect.

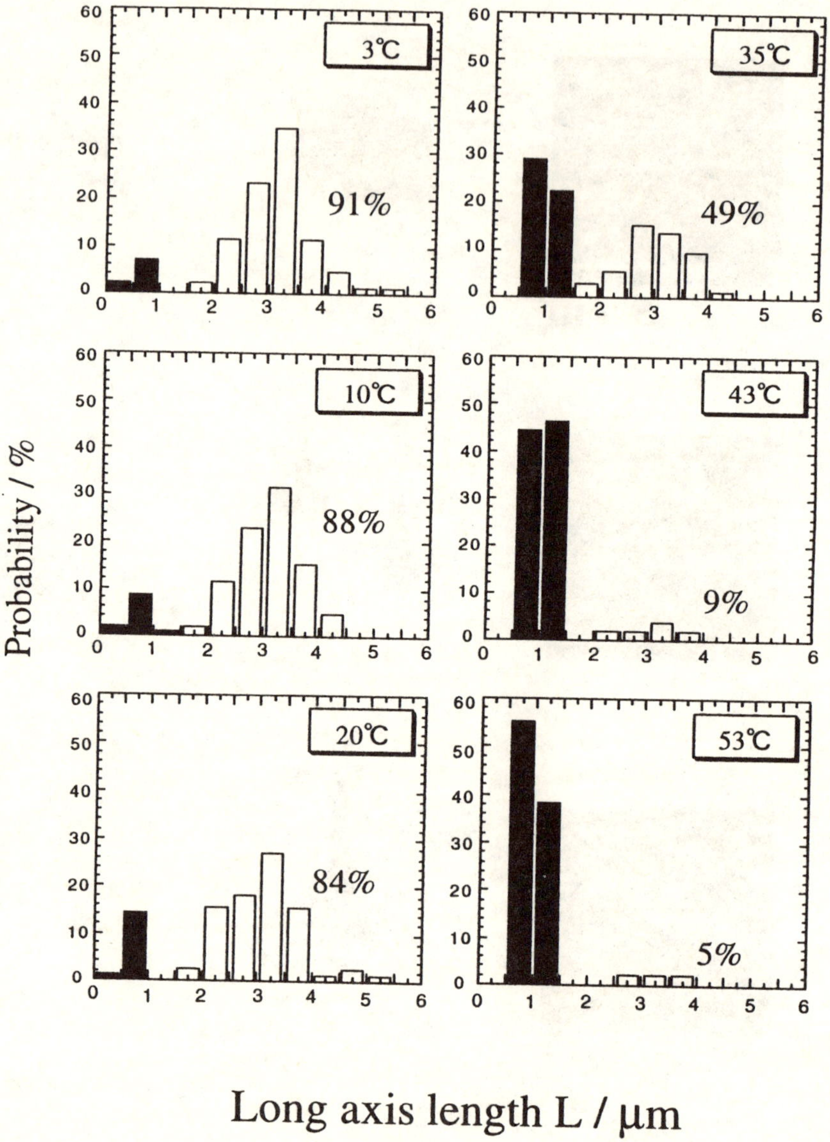

Fig. 3. Histogram of the long-axis length, L, at different temperatures. The filled and open bars indicate globule and coil DNAs, respectively, which were discriminated on the time-successive video images.

such as sodium cation, and the trivalent spermidine will increase accompanied with the increase of the temperature, DNA undergoes the transition from globule to coil.

On the contrary, in the coil-globule transition induced by the addition of neutral polymer chains[11] and also by the addition of alcohol[9], one can expect the transition from coil to globule with the increase of the temperature. This is because the net entropy gain is more preferred in the coil state in the presence of these condensating agents, as is expected from theoretical consideration[11].

Self-Organized Nanostructures from Single Polymer Chain

From the results and discussions given in the preceding sessions, it has become evident that individual polymer chains undergoes first-order phase transition from elongated coil into collapsed globule, if the polymer exhibits enough stiffness and also enough chain length. The experiments with single chain observation clearly indicate that DNA chains fit to such category. It is expected that many of the polyelectrolytes would meet this category when the interval between the charged groups is short enough.

Fig. 4 shows the results of the actual experimental observation on the nanostructured DNA chains[19, 20]. Beside the morphologies of toroid and rod, the formation of spool-like structure is noted. Such spool-like structure was generated with the aid of the core material, as a kind of template, made from cationic lipids. As in the beautiful chromosome structure found in the nuclear in eukaryote, presence of template would be expected to create exotic nanostructures besides the morphologies given in the diagram of Fig.1.

The formation of the phase-segregated state in Fig. 4 is also interesting. As the transition of DNA and possibly of other stiff chains is first-order, the correlation length would not become infinite even at the transition. This means that the intramolecular phase-separation will be generated when the correlation length is small enough. As for DNA, it has been found that the decrease of polarity of the solvent induces the intrachain phase-segregation.

Finally, I would like to indicate the importance of the kinetics on the structure of the collapsed product. In general for condensed matters, it has been well established the kinetic process in first-order phase transition is classified into two different types; nucleation & growth and spinodal decomposition. We have found that single DNA chain actually exhibits nucleation & growth, by the single chain observation with fluorescence microscopy[21]. This result suggests that suitable nonequilibricity is the deterministic factor to generate ordered nanocrystal from single chain through the process of nucleation & growth. If the nonequilibricity exceeds a certain limit, a chain exhibits the route of collapse through spinodal decomposition, indicating that disordered product will be obtained as kinetically frozen state.

Experimental versification of the above mentioned characters in the polymer chains remains as the subject of future investigations by polymer scientists.

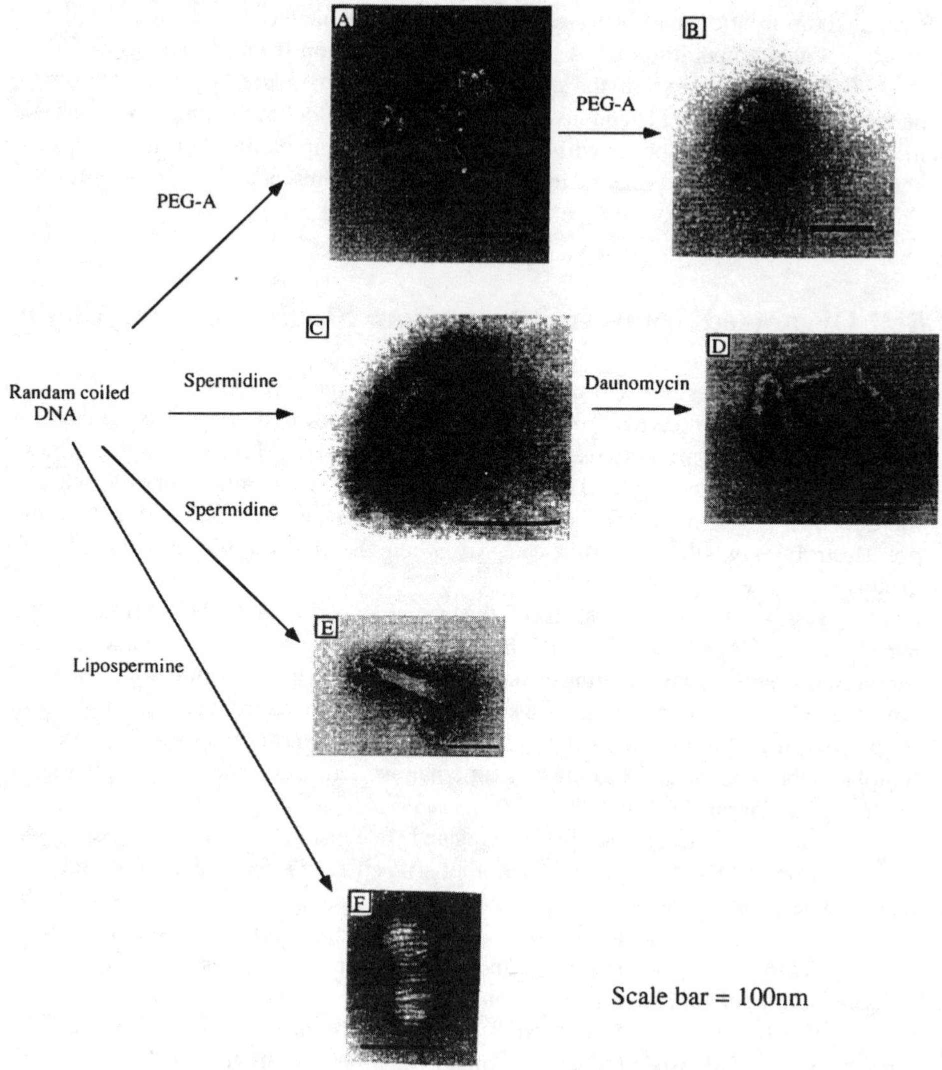

Fig. 4. Morphologies of T4DNAs as were observed with transmission electron microscopy. (Modified from the pictures in Refs. 16, 19, 20, 22)

References

[1] P. J. Flory, *Statistical Mechanics of Chain Molecules* (Interscience, New York, 1969).

[2] P. G. de Gennes, *Scaling Concepts in Polymer Physics* (Cornell University Press,Ithaca, 1979).

[3] A. Yu. Grosberg and A.R. Khokhlov, *Statistical Physics of Macromolecules* (American Institute of Physics, New York, 1994).

[4] V. A. Bloomfield, *Curr. Opin. Struct. Biol.*, **6**, 334-341 (1996).

[5] K. Minagawa, Y. Matsuzawa, K. Yoshikawa, A. R. Khokhlov, and M. Doi, *Biopolymers*, **34**, 555-558 (1994).

[6] K. Yoshikawa, M. Takahashi, V. V. Vasilevskaya, and A. R. Khokhlov, *Phys. Rev. Lett.*, **76**, 3029-3031 (1996).

[7] M.Takahashi, K.Yoshikawa, V.V.Vasilevskaya, and A.R.Khokhlov, *J. Phys. Chem. B*, **101**, 9396-9401 (1997).

[8] Y. Yamasaki and K. Yoshikawa, *J. Am. Chem. Soc.*, **119**, 10573-10578 (1997).

[9] M. Ueda and K. Yoshikawa, *Phys. Rev. Lett.*, **77**, 2133-2136 (1996).

[10] Y. Yoshikawa and K. Yoshikawa, *FEBS Letters*, **361**, 277-281 (1995).

[11] V. V. Vasilcvskaya, A. R. Khokhlov, Y. Matsuzawa, and K. Yoshikawa, *J.Chem. Phys.*, **102**, 6595-6602 (1995).

[12] S. G. Starodoubtsev and K. Yoshikawa, *J. Phys. Chem.*, **100**, 19702-19705 (1996).

[13] S. M. Mel'nikov, V. G. Sergeyev, and K. Yoshikawa, *J. Am. Chem. Soc.*, **117**, 2401-2408 (1995).

[14] S. M. Mel'nikov, V. G. Sergeyev, and K. Yoshikawa, *J. Am. Chem. Soc.*, **117**, 9951-9956 (1995).

[15] H. Noguchi and K. Yoshikawa, *J.Chem. Phys.*, **109**, 5070-5077 (1998).

[16] H. Noguchi, S. Saito, S. Kidoaki, and K. Yoshikawa, *Chem. Phys. Lett.*, **261**, 527-533 (1996).

[17] H.Noguchi and K.Yoshikawa, *Chem. Phys. Lett.*, **278**, 184-188 (1997).

[18] H. Murayama and K. Yoshikawa, submitted for publication.

[19] K. Yoshikawa, Y. Yoshikawa, and Toshio Kanbe, *J. Am. Chem. Soc.*, **119**, 6473-6477 (1997).

[20] Y. Yoshikawa, N. Emi, T. Kanbe, K. Yoshikawa, and H. Saito, *FEBS Lett.*, **396**, 71-76 (1996).

[21] K. Yoshikawa and Y. Matsuzawa, *J. Am. Chem. Soc.*, **118**, 929-930 (1996).

[22] Y. Yoshikawa, K. Yoshikawa, and T. Kanbe, *Biophys. Chem.*, **61**, 93-100 (1996).

Solution Properties of Polymacromonomers Consisting of Polystyrene

Ken Terao, Yo Nakamura, and Takashi Norisuye

Department of Macromolecular Science, Osaka University,
Toyonaka, Osaka 560-0043, Japan
norisuye@chem.sci.osaka-u.ac.jp

Abstract. Analyses are made of *z*-average radii of gyration $\langle S^2\rangle_z^{1/2}$ and intrinsic viscosities $[\eta]$ for two series of polymacromonomer samples consisting only of polystyrene (PS) and having fixed side chain lengths of 15 and 33 styrene monomer units in cyclohexane at 34.5°C (the theta point) and in toluene at 15°C, on the basis of the wormlike chain with or without excluded volume. The current theories for this model are shown to be capable of consistently explaining the main-chain length dependences of $\langle S^2\rangle_z$ and $[\eta]$ for the two polymacromonomers in the theta and good solvents. The chain stiffness is an increasing function of side chain length and higher in the good solvent than in the theta solvent. Thus, the semiflexibility of polymacromonomers under the theta condition arises from the high segment density near the main chain and monomer-monomer repulsions in good solvents further stiffen the polymer backbone.

Introduction

Polymacromonomers are a class of regular comb polymers with the full degree of branching [1]. As was found by a few groups [2-4], those consisting of the poly (methyl methacrylate) (PMMA) backbone and polystyrene (PS) side chains in toluene are much stiffer than linear PMMA and the chain stiffness expressed in terms of the Kuhn segment length λ^{-1} in the Kratky-Porod (KP) wormlike chain [5] becomes higher as the side chain length increases. Undoubtedly, the observed semiflexibility raises a new matter of interest to be explained in terms of the molecular architecture and monomer-monomer interactions, but for the investigation of its origin use of polymacromonomers composed of single polymer species is desirable since interactions between PMMA and PS sub-chains are hard to explore. Furthermore, previous workers [2] encountered a problem that the unperturbed KP model did not always give a consistent explanation of measured *z*-average radii of gyration $\langle S^2\rangle_z$ and intrinsic viscosities $[\eta]$ in toluene.

Thus we recently undertook a series of studies [6-8] on polymacromonomers consisting only of PS to see the applicability of the KP model and effects of solvent (or monomer-monomer interactions) and side chain length on chain stiffness. Figure 1 shows the chemical structure of the polymacromonomer of this kind. We prepared two series of narrow-distribution samples having nearly uniform side chain lengths of 15 and 33 styrene monomer units (designated PMM15 and PMM33, respectively), and determined their second virial coefficients A_2, $<S^2>_z$, and $[\eta]$ in cyclohexane at different temperatures T and in toluene at 15°C as functions of the total weight-average molecular weight M_w. This paper summarizes major findings from the analysis of these data made on the basis of the KP chain with or without excluded volume.

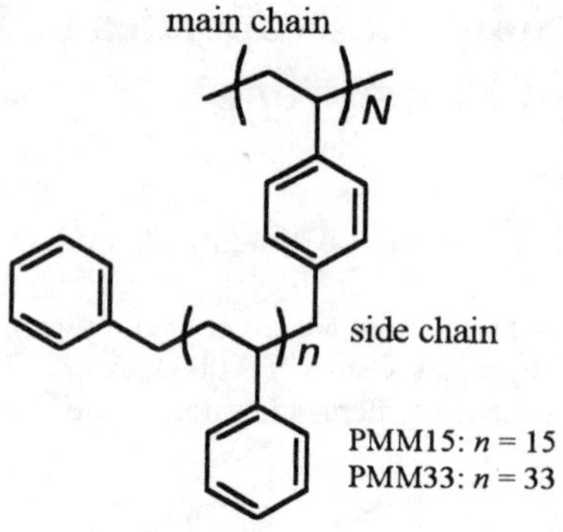

Fig. 1. Chemical structure of a polymacromonomer consisting of polystyrene.

Basic Relations

For convenience of our presentation, we summarize some relevant relations for the KP chain, which is a special case of Yamakawa's general model [9] called the helical wormlike (HW) chain. Excluded-volume effects on $<S^2>_z$ and $[\eta]$ are considered in the quasi-two-parameter (QTP) scheme [9-11].

The unperturbed mean-square radius of gyration $<S^2>_0$ of the KP chain is given by [12]

$$<S^2>_0 = \frac{L}{6\lambda} - \frac{1}{4\lambda^2} + \frac{1}{4\lambda^3 L} - \frac{1}{8\lambda^4 L^2}\left[1 - \exp(-2\lambda L)\right] \tag{1}$$

where L is the contour length of the chain and related to the molecular weight M by $L = M/M_L$, with M_L being the molar mass per unit contour length. If $\lambda L > 2$, this equation can be approximated by [13]

$$\left(\frac{M}{<S^2>_0}\right)^{1/2} = (6\lambda M_L)^{1/2}\left(1 + \frac{3M_L}{4\lambda M}\right) \tag{2}$$

with the maximum error being 1%. The intrinsic viscosity $[\eta]_0$ of the unperturbed chain formulated by Yamakawa and coworkers [14-16] contains one additional parameter, the cylinder diameter d or the bead diameter d_b depending on the hydrodynamic model used for the calculation, but the two theories based on the cylinder [15] and touched-bead [16] models are essentially the same for thin chains with $\lambda d \gtrsim 0.1$ if the small difference between d and d_b are taken into account. On the other hand, for relatively short and thick chains, theoretical $[\eta]_0$ values are available only for the touched-bead model which may not be relevant to hydrodynamically cylindrical polymacromonomers. However, the theories based on the two models reduce to the Einstein equation in the rigid-sphere limit ($L = d =$

d_b) and are equivalent for long chains ($L >> d$ and d_b), so that we may regard $[\eta]_0$ for the touched-bead model chain with $\lambda d_b \gtrsim 0.2$ as that for the cylinder model without significant error and express the latter as

$$[\eta]_0 = f(\lambda L, \lambda d) / (\lambda^3 M) \tag{3}$$

where $f(\lambda L, \lambda d)$ is known [16] as a function of λL and λd (= λd_b).

In the QTP scheme, the radius expansion factor α_s [$\equiv (\langle S^2\rangle / \langle S^2\rangle_0)^{1/2}$] and the viscosity expansion factor α_η [$\equiv ([\eta]/[\eta]_0)^{1/3}$] may be expressed, respectively, by

$$\alpha_s^2 = \left[1 + 10\tilde{z} + \left(\frac{70\pi}{9} + \frac{10}{3}\right)\tilde{z}^2 + 8\pi^{3/2}\tilde{z}^3\right]^{2/15} \times [0.933 + 0.067\exp(-0.85\tilde{z} - 1.39\tilde{z}^2)] \tag{4}$$

$$\alpha_\eta^3 = \left(1 + 3.8\tilde{z} + 1.9\tilde{z}^2\right)^{0.3} \tag{5}$$

if the Domb-Barrett [17] and Barrett [18] functions are adopted. Here, $\tilde{z}$ is the scaled excluded-volume parameter defined by [9]

$$\tilde{z} = \frac{3}{4} K(\lambda L) z \tag{6}$$

with

$$z = \left(\frac{3}{2\pi}\right)^{3/2} (\lambda B)(\lambda L)^{1/2} \tag{7}$$

In eqs 6 and 7, $K(\lambda L)$ is a known function of λL and B is the excluded-volume strength proportional to the familiar binary cluster integral. In sum, $\langle S^2\rangle$ of the KP chain with a given M is determined by λ^{-1}, M_L, and B in the QTP scheme and $[\eta]$, by λ^{-1}, M_L, d, and B.

Data Analysis and Applicability of the KP Model

Our polymacromonomers (PMM15 and 33) attain the theta state in cyclohexane at 34.5°C, as is the case with linear PS [19]. This is shown for three samples of PMM15 in Figure 2, in which A_2 plotted against T is seen to vanish at 34.5°C. In toluene at 15°C, A_2 was positive and large for both PMM15 and 33. Thus, the polymacromonomers resemble linear PS in regard to solubility or solvent power, good or poor.

Figure 3 illustrates the molecular weight dependence of $\langle S^2\rangle_z$ for PMM33 in cyclohexane at the theta point and in toluene at 15°C. The curve for the cyclohexane data (the filled circles) bends slightly down with a slope of 1.4 for M_w

$< 2\times10^6$ and 1.05 for $M_w > 5\times10^6$. This non-Gaussian behavior shows an unmistakable semi-flexibility of the poly-macromonomer. The toluene data appear systematically above the cyclohexane data and follow the indicated solid line whose slope decreases from 1.6 to 1.2 with increasing M_w.

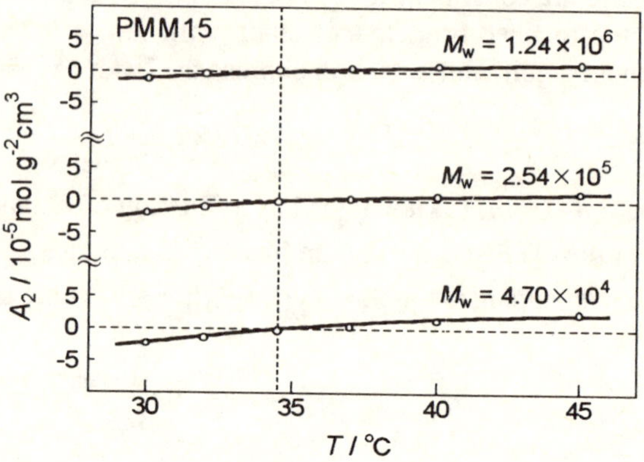

Fig. 2. Temperature dependence of A_2 for three samples of PMM15 in cyclohexane [7]. The vertical dashed line indicates 34.5°C (the Θ temperature).

We determined λ^{-1} and M_L for PMM33 in cyclohexane at 34.5°C to be 22 nm and 13000 nm^{-1}, respectively, from the data in Figure 3 using the linear plot of $(M_w/\langle S^2\rangle_z)^{1/2}$ vs. M_w^{-1} based on eq 2. This plot, when applied to the toluene data, bent down at small M_w^{-1}, indicating that excluded-volume effects are not negligible for $M_w \gtrsim 1.5\times10^6$. Thus these data were analyzed by the combination of eqs 1 and 4, but in actuality, M_L was taken to be 13000 nm^{-1}, the value in cyclohexane.

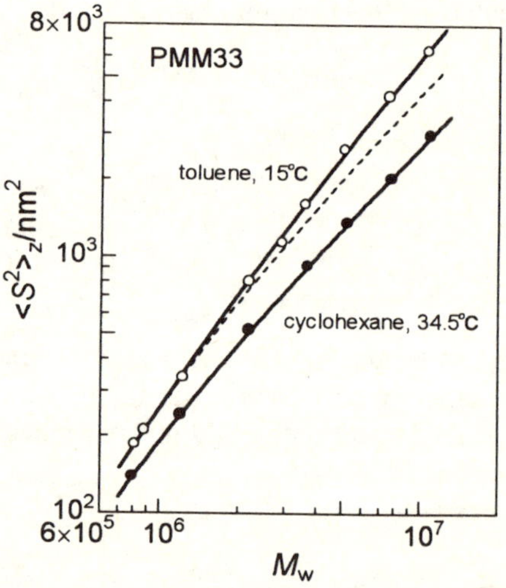

Fig. 3. Molecular weight dependence of $\langle S^2\rangle_z$ for the polymacromonomer PMM33 in cyclohexane at 34.5°C (filled circles) [6] and in toluene at 15°C (unfilled circles) [7]. Solid curves, theoretical values calculated from eqs 1 and 4 with the parameters in Table 1; dashed line, theoretical values for $B = 0$.

The resulting parameters are listed in Table 1, along with those in cyclohexane. The solid curves closely fitting the filled and unfilled circles in Figure 3 actually represent the theoretical values calculated from eqs 1 and 4 with these parameters. The dashed line, which refers to the unperturbed state of the poly-macromonomer in toluene, deviates from the corresponding solid curve progressively with increasing molecular weight, showing that consideration of excluded-volume effects is mandatory in the determination of λ^{-1} and M_L for the polymer in toluene, a good solvent.

Table 1. Model parameters for PMM15 and 33 in cyclohexane and toluene

		from $<S^2>_z$			from $[\eta]$				
Polymer	Solvent	λ^{-1} /nm	$10^{-3}M_L$ /nm^{-1}	B /nm	λ^{-1} /nm	$10^{-3}M_L$ /nm^{-1}	d /nm	B /nm	δ /nm
PMM15	CH [a]	9.5	6.2	0	9.5[c]	6.3	4.7	0	2.2
PMM15	Tol [b]	16	6.2[c]	4.5	16[c]	6.3	5.0	4.5[c]	2.3
PMM33	CH	22	13.0	0	22[c]	13.5	7.4	0	4.0
PMM33	Tol	36	13.0[c]	18	36[c]	13.8	8.5	18[c]	4.3

[a] cyclohexane at 34.5°C; [b] toluene at 15°C; [c] assumed.

The data of $<S^2>_z$ obtained for PMM15 in cyclohexane and toluene were similarly analyzed and found to be quantitatively described by the theories for the KP chain. The estimated parameters are included in Table 1.

The good agreements between theory and experiment found for the two polymacromonomers indicate that possible effects of chain thickness and chain ends on $<S^2>$ are insignificant in the range of M_w studied here. To confirm this, we examined these side chain effects by evaluating the unperturbed mean-square radius of gyration $<S^2>_{0,t}$ of a wormlike polymacromonomer whose side chains are flexible (but yet wormlike) and linked to the stiffer backbone by completely flexible joints. It should be noted that the end effect arises when some side chains near the main-chain ends are more or less oriented to the direction of the main-chain contour, thereby contributing toward apparently increasing L.

The resulting expression is the sum of eq 1 and the contribution $<S^2>_{0,s}$ from the side chains, i.e.,

$$<S^2>_{0,t} = <S^2>_0 + <S^2>_{0,s} \tag{8}$$

$$\begin{aligned} <S^2>_{0,s} = {} & \frac{N-1}{N}\left\{\frac{L_s}{2\lambda_s} - \frac{1}{2\lambda_s^2} + \frac{1}{4\lambda_s^3 L_s}[1-\exp(-2\lambda_s L_s)]\right\} \\ & + \frac{1}{N}\left\{\frac{L_s}{6\lambda_s} - \frac{1}{4\lambda_s^2} + \frac{1}{4\lambda_s^3 L_s} - \frac{1}{8\lambda_s^4 L_s^2}[1-\exp(-2\lambda_s L_s)]\right\} \end{aligned} \tag{9}$$

In eq 9, λ_s and L_s denote the values of λ and L, respectively, for each side chain and N is the number of side chains in the polymacromonomer (i.e., the degree of polymerization of the main chain); note that $L_s = M_s/M_{L,s}$, with M_s and $M_{L,s}$ being the molecular weight and linear mass density of each side chain. The values of M_s determined by light scattering for α-benzyl-ω-hydrogen polystyrene (the precursor) were 1.53×10^3 (for PMM15) and 3.44×10^3 (for PMM33). Substituting these M_s values together with λ_s^{-1} = 2.0 nm and $M_{L,s}$ = 390 nm^{-1} for linear PS [20] into eq 9, we find that the contribution of $<S^2>_{0,s}$ to $<S^2>_{0,t}$ is negligible if $<S^2>_0 \gtrsim$ 70 nm^2 for PMM15 and $<S^2>_0 \gtrsim$ 140 nm^2 for PMM33. The smallest $<S^2>_z$ values determined in this work, 120 nm^2 for the former polymer (not shown here) and 140 nm^2 for the latter polymer (see Figure 3), satisfy these conditions, thus confirming negligible side chain effects on our $\langle S^2 \rangle_z$ data. In passing, eq 8 with eq 9 in the coil limit reduces to the familiar Zimm-Stockmayer equation [21] for Gaussian stars of N arms if L is set equal to zero.

The filled and unfilled circles in panel a of Figure 4 represent the viscosity data for PMM15 samples in cyclohexane at 34.5°C and in toluene at 15°C, respectively. The intrinsic viscosities in both solvents stay almost constant up to $M_w \approx 6\times10^4$ and then increase with increasing molecular weight. The viscosity exponent in the region of $M_w > 4\times10^5$ is 0.48 in cyclohexane and 0.72 in toluene. As panel b of the figure shows, PMM33 exhibits similar viscosity behavior, but the viscosity exponent for $M_w > 10^6$ is 0.59 in cyclohexane and 0.79 in toluene.

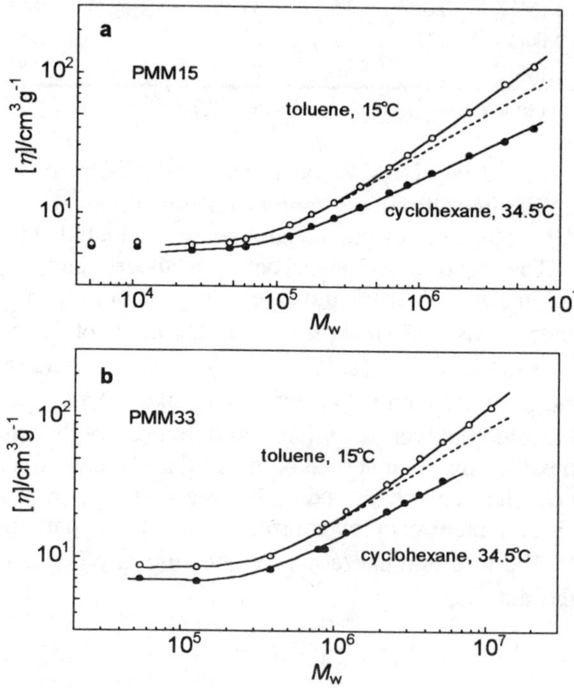

Fig. 4. Molecular weight dependence of $[\eta]$ for the polymacromonomer PMM15 (a) and PMM33 (b) in cyclohexane at 34.5°C (filled circles) and in toluene (unfilled circles) [8]. Solid curves, theoretical values computed from eqs 3 and 5 with eq 10 using the parameters in Table 1; dashed lines, theoretical values for $B = 0$.

With the values of λ^{-1} and B from $\langle S^2 \rangle_z$ (Table 1), we attempted to estimate a set of M_L and d which allows eq 3 with or without eq 5 to fit the $[\eta]$ data for each polymer + solvent system. However, no parameter set explained the very gradual M_w dependence of $[\eta]$ in the low molecular weight region. Since this is most likely due to such an end effect as discussed above, we introduced δ representing the hydrodynamic contribution of side chains (near the main-chain ends) to the contour length and redefined L as

$$L = (M / M_L) + \delta \tag{10}$$

The $[\eta]$ data in Figure 4 were then re-analyzed by eqs 3 and 5 with the λ^{-1} and B values from $\langle S^2 \rangle_z$. The solid curves in the figure are the best-fit theoretical ones for the respective polymer + solvent systems; they are truncated at M corresponding to the Einstein sphere limit (i.e., $L = d$). The agreement between theory and experiment is very satisfactory throughout the molecular weight range in which theoretical $[\eta]_0$ values are available. The dashed lines in the figure again represent the unperturbed theoretical values in toluene.

The model parameters estimated from $[\eta]$ are also summarized in Table 1. The values of M_L are all in good agreement with those determined from $\langle S^2 \rangle_z$ in cyclohexane, substantiating the consistency among the parameters explaining the measured $[\eta]$ and $\langle S^2 \rangle_z$. This finding combined with the observed excellent agreements between theory and experiment allows us to conclude that the KP model is applicable to $\langle S^2 \rangle_z$ and $[\eta]$ of our polymacromonomers in both theta and good solvents.

Geometrical Features of Polymacromonomers

The M_L value of 6250 nm^{-1} (the mean from $\langle S^2 \rangle_z$ and $[\eta]$) for PMM15 with the molar mass 1650 g mol^{-1} of the macromonomer yields a value of 0.26 nm for the monomeric length h along the main-chain contour. Essentially the same h value is obtained for PMM33 whose macromonomer has a molar mass of 3560 g mol^{-1}. This suggests that the local conformation of the PS backbone is insensitive to side chain length. It is of interest to note that the h values of about 0.26 nm are close to that (0.25 nm) calculated on the assumption that the PS backbone assumes the *all-trans* conformation.

The hydrodynamic diameter also provides a piece of information on the poly-macromonomer geometry. The $d/2$ values (2.4 and 3.7 nm) for PMM15 and 33 in cyclohexane are indeed close to the end-to-end distances of 2.4 and 3.9 nm, respectively, calculated for linear PS from the known expression for the unperturbed KP chain with $\lambda_s^{-1} = 2.0$ nm and $M_{L,s} = 390$ nm^{-1} [20]. The values of d in toluene (Table 1) appear somewhat larger than those in cyclohexane. Although this suggests that interactions between neighboring side chains cause the side chains to slightly extend, no definitive conclusion on the solvent effect on d can be drawn from the present limited data.

Effects of Side Chain Length and Solvent on Chain Stiffness

The chain stiffness of linear PS expressed in terms of the Kuhn segment length in the KP chain or the stiffness parameter in the HW chain (a more precise model for the polymer) is known to be about 2 nm [9,20]. The λ^{-1} values of 9.5 nm for PMM15 and 22 nm for PMM33 in cyclohexane at the theta point are much larger than this value, showing that the chain stiffness definitely increases with increasing side chain length, as found by Wintermantel et al. [2,3] for polymacromonomers consisting of the PMMA backbone and PS side chains in toluene, a good solvent.

Our new finding is that even in the absence of intramolecular excluded-volume interaction (in the conventional framework), the semiflexibility of polymacromonomers arises from the high segment density around the main chain.

Though not shown in the present paper, both λ^{-1} and B for our polymacromonomers in cyclohexane very gradually increase with raising temperature [6,7]. Such a correlation between the stiffness and the excluded-volume strength can clearly be seen from Table 1, which shows that the λ^{-1} values for the two polymers in toluene, a good solvent, are 1.6 times as large as those in the theta solvent. Thus, enhancement of repulsion between neighboring side chains and between the main chain and side chain further stiffens the polymacromonomer backbone.

References

1. Y. Tsukahara: in *Macromolecular Design: Concept and Practice* (M. K. Mishra Ed., Polymer Frontiers International Inc., New York, 1994) pp. 161 – 227.
2. M. Wintermantel, M. Schmidt, Y. Tsukahara, K. Kajiwara, and S. Kojima: Macromol. Rapid Commun. **15**, 279 (1994).
3. M. Wintermantel, M. Gerle, K. Fischer, M. Schmidt, I. Wataoka, H. Urakawa, K. Kajiwara, and Y. Tsukahara: Macromolecules **29**, 978 (1996).
4. N. Nemoto, M. Nagai, A. Koike, and S. Okada: Macromolecules **28**, 3854 (1995).
5. O. Kratky and G. Porod: Recl. Trav. Chim. Pays-bas **68**, 1106 (1949).
6. K. Terao, Y. Takeo, M. Tazaki, Y. Nakamura, and T. Norisuye: Polym. J. in press.
7. K. Terao, Y. Nakamura, and T. Norisuye: submitted to Macromolecules.
8. K. Terao, T. Hokajo, Y. Nakamura, and T. Norisuye: to be published.
9. H. Yamakawa: *Helical Wormlike Chains in Polymer Solutions* (Springer, Berlin, 1997).
10. H. Yamakawa and W. H. Stockmayer: J. Chem. Phys. **57**, 2843 (1972).
11. J. Shimada and H. Yamakawa: J. Chem. Phys. **85**, 591 (1986).
12. H. Benoit and P. Doty: J. Phys. Chem. **57**, 958 (1953).
13. H. Murakami, T. Norisuye, and H. Fujita: Macromolecules **13**, 345 (1980).
14. H. Yamakawa and M. Fujii: Macromolecules **7**, 128 (1974).
15. H. Yamakawa and T. Yoshizaki: Macromolecules **13**, 633 (1980).
16. T. Yoshizaki, I. Nitta, and H. Yamakawa: Macromolecules **21**, 165 (1988).
17. C. Domb and A. J. Barrett: Polymer **17**, 179 (1976).
18. A. J. Barrett: Macromolecules **17**, 1566 (1984).
19. See, for example, Y. Nakamura, T. Norisuye, and A. Teramoto: Macromolecules **24**, 4904 (1991).
20. T. Norisuye and H. Fujita: Polym. J. **14**, 143 (1982).
21. B. H. Zimm and W. H. Stockmayer: J. Chem. Phys. **17**, 1301 (1949).

FRICTIONAL BEHAVIOR OF SELF-ASSEMBLED POLYMER BRUSHES

S. M. Kilbey II[†], P. Schorr, M. Tirrell[*]

Department of Chemical Engineering and Materials Science
University of Minnesota
421 Washington Avenue SE
Minneapolis, MN 55455

† current address: Department of Chemical Engineering, Clemson University, Clemson, SC 29634
* author to whom correspondence should be addressed

Abstract. The frictional response between polymer brush layers was studied as a function of solvent condition and degree of compression using a direct force measurement technique. The brush bearing surfaces were made by preferential adsorption of polystyrene-polyvinylpryridine (PS-PVP) block copolymers from a dilute solution. Parallel lateral motion was imparted to one surface, while the response was detected at the opposing brush-covered surface. It was observed, consistent with work published by other groups, that when contacting brushes bathed in a good solvent are sheared past one another, no frictional forces are detected until the polymer layers are strongly compressed. When the solvent is changed to a near-theta solvent, large frictional forces are measured at weaker compressions. The onset of these frictional forces occurs at distances comparable to the distance at which the opposing layers contact one another and rapidly increase as the brushes are compressed. Arguments are advanced that this behavior is attributable to frictional interactions between the polymer chains and the solvent.

Introduction

Many applications for polymers require that they adsorb at interfaces or surfaces. The ability to manipulate the structure and properties of the adsorbing layer provides a route to affect the adherence of materials, the rheology of colloidal suspensions, and the lubricative or wear properties of the surface. Furthermore, utilization of functionalized chains (e.g., telechelic polymers), coupled with the ability to manipulate the molecular-level structure and properties of the tethered layer, offers a potentially attractive means for making microvalves, sensors, or chemo-mechanical layers that are designed to respond to a particular chemical stimuli. In order to develop systems for these applications, it is of critical importance to understand the frictional response of the surface-tethered chains that are subjected to shear due to relative motion of the surrounding fluid medium.

One class of polymers that can be used to confer useful properties to the underlying substrate is diblock copolymers. *A-B* diblock copolymers are extremely effective surface-modifying agents since the amphiphilic nature of these materials allows them to straddle phase boundaries. Manipulating the chemical identity of the polymer segments provides a way to enhance the interfacial activity that each block has with its contacting phase. Controlling the size of each block permits tailoring of the range of interactions across interfaces. *A-B* diblock copolymers are considered to be a model system for studying molecular-level, structure-property relationships of surface-tethered polymer systems due to their ability to form regular, structured layers.

Adsorption of diblock copolymers from a solvent that is good for one block but poor for the other is a well-established route to making so-called "polymer brushes" [1-5]. This scheme of tethering these materials to a substrate is often referred to as preferential or selective adsorption. The block that is not well-solvated will adsorb in a thin layer on a surface, driven by its desire to minimize its contact with the solvent. The adsorption of this "anchor" block tethers the well-solvated, or "buoy" block to the surface. When the surface density is sufficiently high the buoy blocks become crowded. This occurs when the distance between adjacent chains is less than the free-solution radius-of-gyration of the buoy block. To alleviate the lateral crowding, the well-solvated blocks stretch away from the tethering surface, resulting in a brush structure, as shown in Figure 1. Brushes can also be made using triblock copolymers that tether through the center block and with end-functionalized chains. In each case the underlying physics governing brush structure is the same: the osmotic swelling of the brush layer is balanced by the penalty for elastic stretching of the chains. The adsorption behavior and structural features of polymer brushes in various solvent conditions have been extensively studied by both experimental [1-9] and theoretical [10-17] means and are well understood. From this body of work, various strategies have been developed to manipulate the structure of these layers [3, 4, 7].

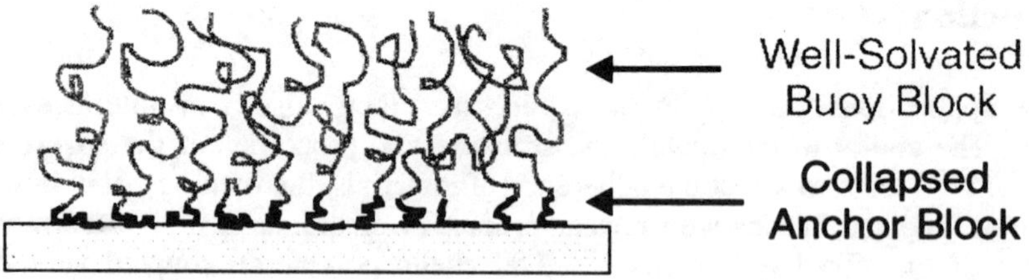

Fig. 1. Schematic drawing of a polymer brush formed by preferential adsorption of *A-B* diblock copolymers.

In this paper we compare the frictional response of a series of block copolymer brushes in both good and near-theta solvents. These experiments were performed using a modified Mark II Surface Forces Apparatus (SFA), which allows the direct measurement of frictional forces between two opposing, brush-bearing surfaces that are in molecular-level contact. When necessary, we refer to "normal" or "structural" force data. These are intimately related to the conformation of the brush chains, and therefore they reflect the structure of the layer. Protocols to self-assemble the brushes and measure the normal forces between the two opposing layers have been well-documented. [1-5,18]

Tailoring the dynamic properties of these systems, however, is an emerging research area. As will be discussed in this paper, the origin of the shear response of brushes in different solvent conditions is not well understood. While there are indications that frictional interactions between the solvent and segments of the brush control the strength of the shear response [18, 19], the origin of the molecular-level shear properties of polymer brushes is a subject of much debate [18-29]. Reference 29 provides a concise review of this subject.

Experimental

Materials. The di- and triblock copolymers used in these experiments are listed in Table 1. These polymers were anionically polymerized, the details of which are given elsewhere [3], and the polydispersity index of each was less than 1.08.

Table 1. Summary of the PS-PVP block copolymers used in this study. As noted in the table, one PS-PVP-PS triblock copolymer, which tethers two PS arms to the surface when the center PVP block adsorbs, was examined.

Polymer	M_{PVP}	M_{PS}
[45-55]k	45,000	55,000
[15-90]k	15,000	92,000
[70-70]k	70,000	70,000
[39-102-62]k	102,000	38,000 / 62,000
[103-114]k	103,000	114,000

Procedure. All experimental work was carried out at 32 ± 0.1 °C. Brushes were self-assembled on the opposing surfaces of the SFA from a dilute toluene solution. This creates a PS brush that is tethered to the surface through its PVP block, which is not solvated by toluene. Normal forces were measured as a function of the separation distance between the brush-bearing surfaces while the layers were bathed in toluene. After measurement of the normal forces, frictional forces were measured at several surface-separation distances. At each separation distance a few different frequencies and amplitudes were explored [30]. The modifications to the SFA that enable molecular-level shear motion and detection have been described in a previous publication [31]. In the shearing experiments described in this paper, a triangular ramp signal was used to drive one surface past the other. Following the friction measurements in the good solvent, the solvent was changed to cyclohexane, which is a near-theta-solvent for PS brushes at this temperature (T_{theta}, PS = 34.5 °C). Normal force and friction measurements were then recorded in the near-theta solvent. Following these experiments, the solvent was changed back to toluene and the normal forces were again measured to re-examine the structure of the brush layers.

Results and Discussion

Figure 2 shows the normal and frictional forces for the (103-114)k PS-PVP diblock copolymer brush as a function of surface separation distance in toluene and cyclohexane. For clarity, lines have been fit to normal force profiles to guide the eye. As seen in the Figure 2, the range of the normal forces for the brush in toluene is longer-ranged than that of the brush in cyclohexane because the polystyrene block is more swollen in the good solvent. Since there is a brush on each of the opposing surfaces, the distance at which the structural forces becomes non-zero is approximately twice the height of the brush layer. The brushes studied in these experiments are swollen 1.2 to 2.1 times further in toluene than in the near-theta solvent, cyclohexane. This agrees well with the theoretical prediction of Halperin [14]. Finally, we note that in each solvent, the normal forces between the layers are monotonically repulsive, indicating that the brush segments repel each other when the brushes are brought into contact. Aside from the differences in the range of normal force interactions due to the solvent condition, brushes exhibit similar structural behavior in both solvents. The dynamic behavior in the two solvent conditions, however, is strikingly different. In cyclohexane (near-theta condition), the onset of frictional forces occurs at a separation distance comparable to the onset of normal forces, while in toluene (good solvent condition), frictional forces are not observed until the layers are highly compressed.

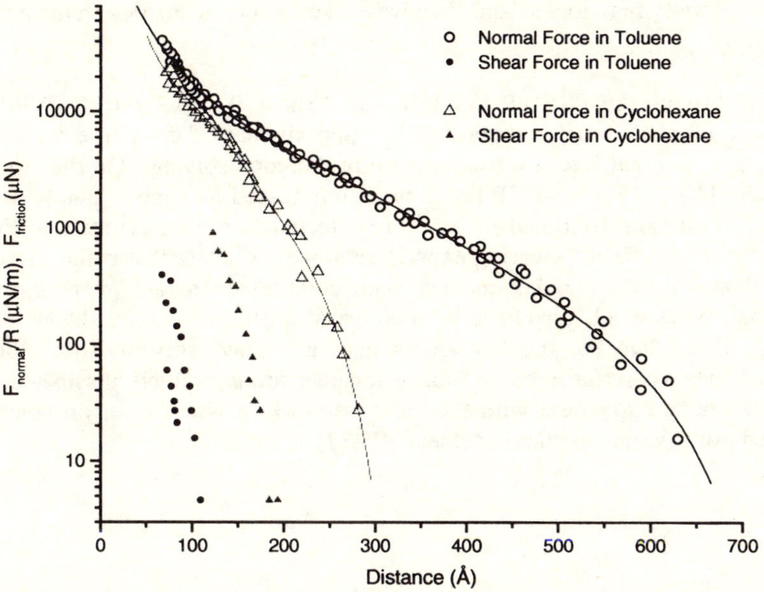

Fig. 2. Normal and shear forces between two opposing PS-PVP [103-114]k brushes in both good (toluene - open and filled circles) and near-theta (cyclohexane - open and filled triangles) solvents. In both cases, normal force data was obtained before and after the shearing motion was applied to check for damage to the underlying brush structure. The reproducibility of the normal force data demonstrated that no desorption of polymer chains or damage to the layer took place during the frictional measurements. The applied shear velocity was 50 nm/sec. The solid lines are drawn to guide the eye.

When the brushes were immersed in toluene and brought into contact, no frictional forces were transmitted between the layers. As the brushes were compressed against one another, frictional forces continued to be dissipated effectively. This can be easily seen for all of the brushes studied in Figure 3. In this figure, the frictional force data are plotted as a function of the gap thickness (D) normalized by the separation distance at which normal forces were detected (equal to twice the equilibrium brush height, $2L_o$). It is worth noting that a frictional force of zero does not imply a frictionless condition, but rather that any force transmitted between the layers was below the detection limit of the device (~ 1 μN). Despite the bimodal distribution of PS chain lengths that make up the brush formed by the PS-PVP-PS triblock copolymer, the shearing results exhibit the same pattern of behavior exhibited by the diblock copolymer brushes. This is due to the fact that compression of the bimodal layer tends to homogenize the segment density distribution in the layer -- in essence, a compressed brush made from a triblock copolymer looks and behaves like a brush made from a diblock copolymer.

Even through the PS-PVP [55-45]k, [90-15]k, [70-70]k, and PS-PVP-PS [39-102-62]k brushes were compressed to approximately 1/5 of their equilibrium height, no frictional forces were observed in the good solvent. On the other hand, when the [103-114]k PS-PVP brush was compressed to approximately 1/6 of its equilibrium height, frictional forces were detected as seen previously in Figure 2. For these good solvent shearing experiments, we speculate that if the triblock and other three diblock copolymers had been compressed to a higher degree, then frictional forces would also have been observed for these systems. Nevertheless, it is clear that when the brushes are bathed in a good solvent, frictional forces between the two surfaces being rubbed together are effectively dissipated. These findings are in agreement with the previous work of Klein *et al* on similar end-tethered polystyrene brushes in toluene [28,32].

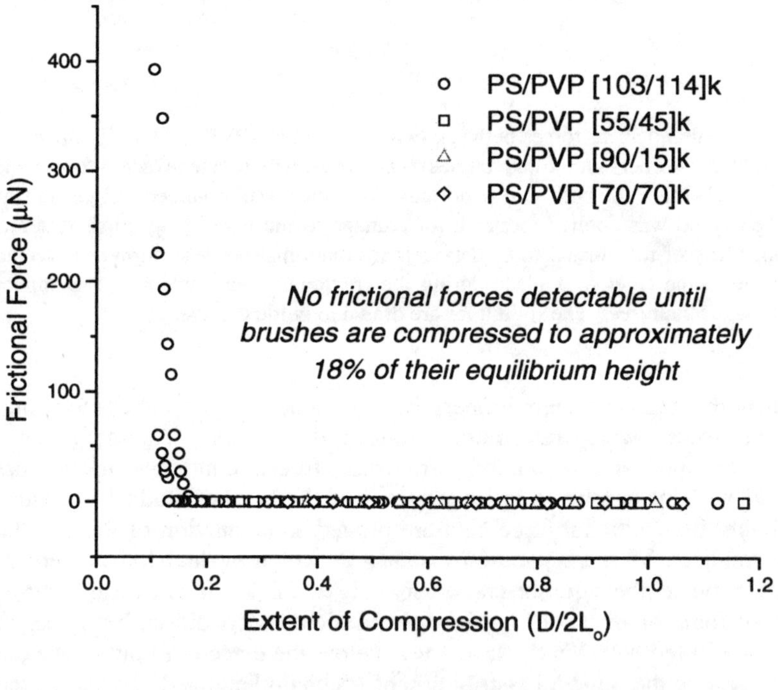

Fig. 3. Frictional forces between PS-PVP polymer brushes in toluene (good solvent). The separation distance (D) has been scaled by twice the equilibrium brush height of the different polymer brushes investigated ($2L_o$).

Shearing brushes in the near-theta solvent, however, shows a very different pattern of behavior. As shown in Figure 4 for all of the brushes studied, frictional forces were detected at a distance comparable to the onset of normal force repulsion in the same solvent. This indicates that as soon as the brush layers contact each other, shear forces are transmitted across the interface, which is in stark contrast with the frictional behavior of the well-solvated brushes. These frictional forces increase quickly as the layers are compressed. It can be seen from the data in Figure 4 that there is no apparent dependence of the frictional forces on molecular weight of the PS block. Also, no apparent dependence on the surface density of PS chains was seen [18].

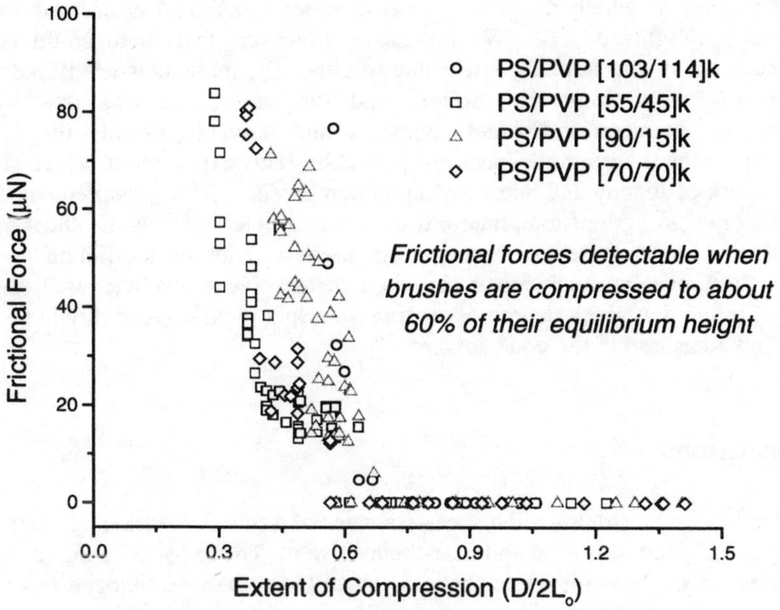

Fig. 4. Frictional forces between PS-PVP polymer brushes in cyclohexane (near-theta solvent). The separation distance (D) has been scaled by twice the equilibrium brush height of the different polymer brushes investigated ($2L_o$).

While these two factors, molecular weight and tethering density, control the structure of the layer, they do not appear to be the controlling influence in the frictional behavior of these systems. Several other effects have been postulated, including mutual interpenetration of chains from opposing brushes [25-27], "freezing" of the chains to a glassy state [28], and frictional interactions between the segments and the solvent [18]. The idea that the pattern of behavior seen in both solvents can be ascribed to the interpenetration and entanglement between the brushes seems to be at odds with numerical self-consistent field (SCF) by Hirz [12]. These SCF calculations, which model a brush with a tethering density and molecular weight similar to the brushes studied in our experiments, showed that opposing brushes at similar levels of compression interpenetrate each other to similar depths whether they are immersed in a good or theta solvent. It remains an open issue, however, as to whether the shearing action affects the brush structure enough so that interpenetration becomes favorable, or whether surface rearrangement of the layer [33] could lead to the differences in the pattern of behavior seen in the good and near-theta solvent cases. And while confinement on molecular length scales has been known to lead to freezing of molecules [34], the concentration (expressed as the mass of PS segments in the volume between the confining surfaces) to which the brushes are squeezed remains below the concentration at which the ODT (order-disorder transition) is crossed for these molecular weights of PS. We recognize, however, that there could be some enhancement of the freezing effect due to either the molecular-level confinement or heterogeneity along the surface. At this stage, the idea that frictional interactions between individual segments and solvent controls the frictional behavior of these systems seems most plausible. This explanation, which is related to the work of Joanny and Semenov on molten brushes [35,36], is predicated on the idea that the strength of monomeric friction per contact controls the shear response of solvated brush systems. Since the monomeric friction coefficient in a good solvent (toluene) is much lower than in a theta solvent (cyclohexane), the shear force measured at given degree of compression in a theta solvent should be greater than that measured in the good solvent.

Conclusions

Changes in the structure and shear-response of a few PS brushes of various sizes were investigated in a good and near-theta solvent. The expected contraction in the thickness of the brush layer is observed when the solvent is changed from a good solvent to a near-theta solvent. This collapse of the brush chains is due to the decrease in the osmotic force that swells the layer. Despite the change in solvent quality, the structural forces between the layer remain repulsive.

When two brushes are laterally sheared past each other in a good solvent, shear forces are not detected until the brushes are strongly compressed. This indicates that the brushes are effective in dissipating the shear stress across the interface between the opposing layers. In the near-theta solvent, however, frictional forces are seen when the brushes come into contact with one another, and the strength of these frictional forces increases quickly as the layers are compressed. This pattern of behavior is attributed to the strength of frictional interactions between the brush segments and solvent. While this hypothesis can account for the difference in the pattern of behavior seen between brushes sheared in good and near-theta solvents, shear-induced stretching or swelling of the brush layer and surface-rearrangement of the chains remain as possibilities that could also contribute to the observed pattern of behavior.

References

1. Hadziioannou, G., S. Patel, M. Tirrell, *J. Am. Chem. Soc.* **108**, 2869 (1986).
2. Tirrell, M., S. Patel, G. Hadziioannou, *Proc. Natl. Acad. Sci.* **84**, 4725 (1987).
3. Watanabe, H., M. Tirrell, *Macromolecules* **26**, 6455 (1993).
4. Dhoot, S., H. Watanabe, M. Tirrell, *Colloids and Surfaces A* **86**, 47 (1994).
5. Patel, S.S., M. Tirrell, *Annu. Rev. Phys. Chem.* **40**, 597 (1989).
6. Auroy, P., Y. Mir, L. Auvray, *Phys. Rev. Lett.* **69**, 93 (1992).
7. Karim, A., S.K. Satija, J.F. Douglas, J.F. Anker, L.J. Fetters, *Phys. Rev. Lett.* **73**, 3407 (1994).
8. Mansfield, T.L., D.R. Iyengar, G. Beaucage, T.J. McCarthy, R.S. Stein, RJ. Composto, *Macromolecules* **28**, 492 (1995).
9. Webber, R.M., C.C. van der Linden, J.L. Anderson, *Langmuir* **12**, 1040 (1996).
10. Alexander, S, *J. Phys. (France)* **38**, 983 (1977).
11. de Gennes, P.G., *Macromolecules* **13**, 1069 (1980).
12. Hirz, S.J., Masters Thesis, University of Minnesota (1986).
13. Milner, S.T., T.A. Witten, M.E. Cates, *Macromolecules* **21**, 2610 (1988).
14. Halperin, A, *J. Phys. (France)* **49**, 547 (1988).
15. Dolan, A.K., S.F. Edwards, *Proc. R. Soc. Lond. A* **343**, 427 (1975).
16. Cosgrove, T., T. Heath, B. van Lent, F. Leermakers, J.H.M.S. Scheutjens, *Macromolecules* **20**, 1692 (1987).
17. Milner, S.T., , *J. Chem. Soc. Faraday Trans.* **86**, 1349 (1990).
18. Kilbey II, S.M., Doctoral Thesis, University of Minnesota (1996).
19. Kilbey II, S.M., F.S. Bates, M. Tirrell, unpublished data.
20. Barrat, J.-L, *Macromolecules* **25**, 832 (1992).
21. Rabin, Y., S. Alexander, *Europhys. Lett.* **13**, 49 (1990).
22. Miao, L., H. Guo, M.J. Zuckerman, *Macromolecules* **29**, 2289 (1996).
23. Buguin, A., F. Brochard-Wyart, *Macromolecules* **29**, 4937 (1996).
24. Klein, J., E. Kumacheva, D. Perahia, D. Mahalu, S. Warburg, *Faraday Discuss.* **98**, 173 (1994).
25. Dhinojwala, A., L. Cai, S. Granick, *Langmuir* **12**, 4537 (1996).
26. Granick, S., A.L. Demirel, L.L. Cai, J. Peanasky, *J. Chem. (Israel)* **35**, 75 (1995).
27. Cai, L.L., J. Peanasky, S. Granick, *Trends in Polym. Sci.* **4**, 47 (1996).
28. Klein, J., *Annual Review of Material Science* **26**, 581 (1996).
29. Grest, G., to appear in *Adv. Polym. Science*.
30. Shearing velocities used ranged from 4.4 nm s-1 to 70 nm s-1 and the maximum amplitude used was less than 2 μm. During the shearing experiments, the movement of the surfaces normal to one another was less than 10 Å at the largest shear amplitudes explored.
31. Kilbey II, S.M., Bates, F., Tirrell, M., Yoshizawa, H., Hill, R., Israelachvili, J., *Macromolecules* **28**, 5626 (1995).
32. Klein, J., Kumacheva, E., Mahalu, D., Perahia, D., Fetters, L.J., *Nature*, **370**, 634 (1994).
33. Kelley, T.W., Schorr, P., Johnson, K.D., Tirrell, M., Frisbie, D., *Macromolecules*, **31**, 4297 (1998)
34. see for example, Israelachvili, J.N, *Surf. Sci. Reports* **14**, 109 (1992).
35. Joanny, J.-F., *Langmuir* **8**, 989 (1992).
36. Semenov, A.N, *Langmuir* **11**, 3560 (1995).

Printing: COLOR-DRUCK DORFI GmbH, Berlin
Binding: Buchbinderei Lüderitz & Bauer, Berlin